SPECIALTY PAPERS

SPECIALTY PAPERS

Francis R. Gould

NOYES DATA CORPORATION
Park Ridge, New Jersey, U.S.A.
1976

Library of Congress Catalog Card Number: 76-24144
ISBN: 0-8155-0639-2
Printed in the United States

Published in the United States of America by
Noyes Data Corporation
Noyes Building, Park Ridge, New Jersey 07656

FOREWORD

The detailed, descriptive information in this book is based on U.S. patents issued since 1968 that deal with specialty papers. This includes one reissue.

This book serves a double purpose in that it supplies detailed technical information and can be used as a guide to the U.S. patent literature in this field. By indicating all the information that is significant, and eliminating legal jargon and juristic phraseology, this book presents an advanced, technically oriented review of the manufacture and use of specialty papers as depicted in U.S. patents.

The U.S. patent literature is the largest and most comprehensive collection of technical information in the world. There is more practical, commercial, timely process information assembled here than is available from any other source. The technical information obtained from a patent is extremely reliable and comprehensive; sufficient information must be included to avoid rejection for "insufficient disclosure." These patents include practically all of those issued on the subject in the United States during the period under review; there has been no bias in the selection of patents for inclusion.

The patent literature covers a substantial amount of information not available in the journal literature. The patent literature is a prime source of basic commercially useful information. This information is overlooked by those who rely primarily on the periodical journal literature. It is realized that there is a lag between a patent application on a new process development and the granting of a patent, but it is felt that this may roughly parallel or even anticipate the lag in putting that development into commercial practice.

Many of these patents are being utilized commercially. Whether used or not, they offer opportunities for technological transfer. Also, a major purpose of this book is to describe the number of technical possibilities available, which may open up profitable areas of research and development. The information contained in this book will allow you to establish a sound background before launching into research in this field.

Advanced composition and production methods developed by Noyes Data are employed to bring our new durably bound books to you in a minimum of time. Special techniques are used to close the gap between "manuscript" and "completed book." Industrial technology is progressing so rapidly that time-honored, conventional typesetting, binding and shipping methods are no longer suitable. We have bypassed the delays in the conventional book publishing cycle and provide the user with an effective and convenient means of reviewing up-to-date information in depth.

The Table of Contents is organized in such a way as to serve as a subject index. Other indexes by inventor, patent number and company help in providing easy access to the information contained in this book.

15 Reasons Why the U.S. Patent Office Literature Is Important to You –

1. The U.S. patent literature is the largest and most comprehensive collection of technical information in the world. There is more practical commercial process information assembled here than is available from any other source.

2. The technical information obtained from the patent literature is extremely comprehensive; sufficient information must be included to avoid rejection for "insufficient disclosure."

3. The patent literature is a prime source of basic commercially utilizable information. This information is overlooked by those who rely primarily on the periodical journal literature.

4. An important feature of the patent literature is that it can serve to avoid duplication of research and development.

5. Patents, unlike periodical literature, are bound by definition to contain new information, data and ideas.

6. It can serve as a source of new ideas in a different but related field, and may be outside the patent protection offered the original invention.

7. Since claims are narrowly defined, much valuable information is included that may be outside the legal protection afforded by the claims.

8. Patents discuss the difficulties associated with previous research, development or production techniques, and offer a specific method of overcoming problems. This gives clues to current process information that has not been published in periodicals or books.

9. Can aid in process design by providing a selection of alternate techniques. A powerful research and engineering tool.

10. Obtain licenses – many U.S. chemical patents have not been developed commercially.

11. Patents provide an excellent starting point for the next investigator.

12. Frequently, innovations derived from research are first disclosed in the patent literature, prior to coverage in the periodical literature.

13. Patents offer a most valuable method of keeping abreast of latest technologies, serving an individual's own "current awareness" program.

14. Copies of U.S. patents are easily obtained from the U.S. Patent Office at 50¢ a copy.

15. It is a creative source of ideas for those with imagination.

CONTENTS AND SUBJECT INDEX

INTRODUCTION

Paper is used in a wide variety of forms, at home, at school, in business and in industry. Americans utilize about 500 pounds of paper per person annually. The pulp and paper industry ranks tenth among U.S. manufacturing industries in the value of its shipments.

The initial step in papermaking, the digestion of wood chips into fiber was reported in the first book in this series *Pulp Mill Processes—Pulping, Bleaching, Recycling.* The next step in the process, the conversion of the fibrous pulp into paper, and the use of different wet end additives for imparting desired properties to paper was described in *Paper Manufacture.*

The present volume is concerned with the manufacture of special papers, largely designed for specific home or industrial applications. Included are decorative, security, photographic, extensible, erasable, absorbent, electroconductive, and insulating papers, filters, papers for food products and paper protected against microbial deterioration.

PRESERVATION OF PAPER

PROTECTION AGAINST FUNGUS AND BACTERIA

$CuSO_4$-8-Hydroxyquinoline Sulfate-H_2SO_4 Paper Fungicide

The process developed by *B.O. Hager; U.S. Patent 3,713,963; January 30, 1973; assigned to Hager Aktiebolag, Sweden* relates to means for protecting fibrous organic materials, for example, paper, board, etc., against destructive organisms such as fungi or insects.

The preservative can in its simplest form be composed as a mixture of copper sulfate and a reaction product of 8-hydroxyquinoline with a strong acid, for example, 8-hydroxyquinoline sulfate. These two substances are easy to dissolve, and it has been proven possible to mix them in dry state without having a reaction between them.

In order to diminish the risk for such reaction in the mixture, and in order to facilitate the dissolution (in the aqueous suspension of pulp) of the material some sulfuric acid may be added. The preservative obtained in this way is a non-dusting powder, easy to use. For instance, it can be given the following composition (stated in parts by weight).

	Parts by Weight
Copper sulfate, crystallized	85
8-Hydroxyquinoline sulfate	10
Sulfuric acid, 65%	5

If arsenic acid is included in the preservative, the preservative may, for example, be given the following composition.

	Parts by Weight
Copper sulfate, crystallized	60
Copper sulfate, water-free	10

(continued)

	Parts by Weight
8-Hydroxyquinoline sulfate	10
Monosodium arsenate, crystallized	15
Sulfuric acid, 65%	5

In this latter composition some water-free copper sulfate has been used in the mixture for the reason that the powder hereby acquires a more suitable form for its handling.

In the manufacture of the fiber product, for instance paper, the easily dissoluble preservative is introduced into the paper raw material. As a rule, it is added to the mixture of paper pulp and water in the pulp engine. The amount of paper pulp here usually is about 3%. The amount of preservative that ought to be added to the paper pulp depends on the effect of the preservation wanted and on the particular paper pulp being treated, different pulps having different retentions.

The amount of preservative added varies normally between 0.5 and 3.0% by weight based on the dry weight of the pulp. By the addition of the preservative to the pulp-water mixture the different compounds in the preservative are dissolved but in the moment of dissolution they form a voluminous precipitation which is deposited on the fibers.

In order to obtain best retention, the acidity of the pulp-preservative-water mixture has to be adjusted to a suitable value between pH 6.2 and 7.0. Retentions around 70% of theoretical are obtained with proper handling and suitable pulp. Good results have been obtained with unbleached sulfate pulp. The retention is favored when the pulp has a high chlorine value, that is, when it also contains some lignin, resin and similar substances in addition to the cellulose fibers per se.

The proportions of the ingredients of the preserving composition can be varied if wanted or necessary. Other active metals than copper can be used in the preservative, for instance, zinc, nickel, cadmium, etc., but, as stated above, the copper radical is as a rule the most suitable. Metals such as aluminum and/or chromium can be introduced in the preservative in the form of soluble salts in order to serve as an additional fixing agent, particularly where arsenic acid is the substance to be fixed.

Waste from plants using preservatives of the process can be cleaned (a) by a filtration after an addition of alkali, or (b) by a filtration through slaked lime. Residues of the preservatives may even be removed (c) by passing the waste through finely divided iron.

Moldproof Paper Containing Metallic Quinolinolates

A.R. Bowers and D.T. Milne; U.S. Patent 3,493,464; February 3, 1970; assigned to Mosinee Paper Mills Company describes a fungus resistant or moldproof paper containing a metallic quinolinolate as the fungicidal or fungistatic agent which has been precipitated in the pulp slurry by first adding a solution of a hydroxyquinoline to the pulp slurry and then adding a solution of a metal salt, to thereby precipitate the metallic quinolinolate prior to formation of a sheet from the pulp slurry or furnish. The hydroxyquinoline is 8-hydroxyquinoline, 2-hydroxyquinoline, dibromo-8-hydroxyquinoline, dichloro-8-hydroxyquinoline, a salt of one of the preceding hydroxyquinolines, or a mixture of at least two of the hydroxyquinolines or salts. The metallic salt is a salt of copper, zinc, aluminum, cobalt, manganese, nickel, iron, or a mixture of at least two of the foregoing.

The paper can be formed as a flat, creped or machine glazed sheet, it can be bleached, unbleached or colored in any desired manner, it may be sized or unsized, and it may or may not contain a wet strength agent, all without impairment of its mold resistance properties.

The term "parts" as used in the examples refers to parts by weight on the basis of 100 parts of oven-dry fibers or pulp, unless otherwise noted. "Basis weight" is the weight in pounds of 3,000 square feet of paper. All papers were tested for fungus resistance in accordance with TAPPI (Technical Association of the Pulp and Paper Industry) Method T 487 m-54 using five test organisms: *Chaetomium globosum, Aspergillus niger, Aspergillus terreus, Aspergillus fumigatus,* and Aspergillus species (unidentified), unless otherwise noted.

The soil burial tests utilized in the examples are in accordance with the method of ASTM (American Society for Testing Materials) E154-64T, except that a soil bed eighteen inches deep and four feet square was used. Briefly, the soil bed was maintained at a temperature of 83°F (plus or minus 2°), and a relative humidity of 73 to 75%. Test samples of the paper were inserted in plastic frames and buried in the soil bed for the times specified in the examples.

The term "deterioration" used in the examples to characterize results of the soil burial tests means the disintegration or rotting away of the paper due to the action of soil fungi and soil bacteria. When the paper is to be used as a mulch paper, it is essential that it not deteriorate throughout the growing season for the particular crop with which it is used; hence, resistance to soil deterioration for useful periods of time is a necessary characteristic of a suitable mulch paper.

The following mixing procedure was used. The hydroxyquinoline was first dissolved in an acid or alkaline solution and added to a well-agitated aqueous slurry of the fibers in either the pulper or beater. The mixture was agitated so that the hydroxyquinoline became well soaked into the fibers; suitable results are obtained with contact times of about 2 minutes to 2 hours depending on the nature of the mixing in the beater or pulper and the quantity of pulp, with a contact time of about 10 minutes used in most of the examples.

Thereafter, a metal salt solution was added to the slurry to precipitate a metallic quinolinolate in and among the fibers, after which other ingredients, such as size, retention aid, wet strength resin, etc. were added to the slurry or furnish. The furnish was formed into paper on a conventional Fourdrinier papermaking machine, although this process may also be employed with cylinder papermaking machines.

Example 1: Fungus Resistant Soap Wrap – To an aqueous slurry comprising 80 parts full-bleached softwood kraft fibers and 20 parts bleached hardwood kraft fibers were added, after dissolving outside the pulper, 0.3 part 8-hydroxyquinoline, 0.07 part sodium hydroxide and 0.27 part copper sulfate pentahydrate. The 8-hydroxyquinoline was dissolved with the sodium hydroxide before addition to the slurry, and the copper sulfate was added about 10 minutes after the addition of the 8-hydroxyquinoline and sodium hydroxide.

Other additions were 5 parts china clay, 1.0 part modified rosin sizing, 0.3 part urea-formaldehyde wet strength resin, 2.0 parts alum to pH 4.5 and 0.75 part of a high molecular weight cationic polyamine bisulfate salt.

This furnish was formed into a dried sheet on a Fourdrinier paper machine with a Yankee dryer. After normal drying, the paper contained 0.17 part copper 8-quinolinolate for every 100 parts of fibers, and had a basis weight of 21.9 pounds and a caliper of 1.75 mils. The machine glazed sheet had a Sheffield smoothness of 60.

Samples were leached for 24 hours in standing tap water and were then tested for fungus resistance. The leached samples completely inhibited the growth of the test organisms for the two week incubation period, but control sheets without the copper 8-quinolinolate showed prolific fungus growth. Paper of the basis weight and having the copper 8-quinolinolate quantity of this example provides a fungus resistant paper suitable for soap wrap.

Example 2: Fungus Resistant Mulch Paper – A solution of 0.98 part of 8-hydroxyquinoline dissolved with 0.25 part of sodium hydroxide in 98.85 parts of water was added to an aqueous slurry comprising 100 parts of natural kraft softwood fibers and approximately 3,300 parts of water in a pulper. After thorough mixing in the pulper for ten minutes, a solution of 0.88 part of copper sulfate pentahydrate in 99.12 parts of water was added to the pulper to form 1.19 parts by weight of copper 8-quinolinolate in the slurry.

After thorough mixing to insure complete precipitation of the copper 8-quinolinolate, 1.875 parts of carbon black and 3.5 parts of papermaker's alum were added and mixed thoroughly with the fibrous slurry. The slurry was hydrated to a Canadian Standard Freeness of 450 to 525 cc by passing through conventional Jordan engines.

Rosin sizing was added continuously to the hydrated slurry at a rate of 1.2 parts for 100 parts by weight of kraft fibers and urea-formaldehyde wet strength resin was added continuously at the rate of 1.0 parts of resin for every 100 parts by weight of kraft fibers. After dilution to a consistency of 0.3 to 0.6 part of fibers to 100 parts of water a high molecular weight polyacrylamide retention aid was added at the rate of 0.025 part by weight to 100 parts by weight of kraft fibers. The furnish was formed into a paper sheet on a Fourdrinier paper machine. After pressing and partial drying, the sheet was creped in conventional manner and dried to a moisture content of from 3 to 6%.

The sheet, so formed, was black in color, had a basis weight of 50.4 pounds, and was 0.0105 inch thick. Its machine direction breaking stretch was from 3 to 5%. The sheet contained from 0.96 to 1.05 parts of copper 8-quinolinolate for every 100 parts of fibers.

The light transmittance of this paper in direct sunlight was less than 5 foot candles. This is more than adequate to suppress all weed growth. The black color serves to warm the soil over which the paper is laid, with resulting faster seed germination and plant growth. The sheet was tested for resistance to fungal attack, according to the above procedure, and, at the end of the two week test period, was found to be completely free from growth of the five test organisms. Paper from the same furnish but without copper 8-quinolinolate was subjected to the same tests, and exhibited heavy growth of all organisms at the end of one week and at the end of two weeks. The paper also was subjected to the soil burial test, and there was no visible deterioration of the sheets at the end of 1, 2, 3 and 4 weeks. The tensile strengths of the sheets were determined and found to be as follows.

	Lb/In Width	
Burial Time	CD*	MD**
Unexposed	11.3	14.7
One week	11.1	14.8
Two weeks	11.9	14.1
Three weeks	11.5	13.9
Four weeks	10.4	13.4

*Cross directional
**Machine directional

Untreated paper (that is, from the same furnish but without the precipitated copper 8-quinolinolate) of the same basis weight was subjected to the same soil burial test at the same time, and was found to have decomposed completely at the end of one week.

The paper withstood mechanical laying, perforation, and seeding operations when applied in a field. The paper remained in the field in the Mississippi Delta and suppressed weed growth until harvesting of the crop, which took place from 90 to 120 days after planting. After harvesting the crop, it was possible to disk the paper into the soil during preparation for the next crop.

Example 3: Polymer Coated Paper – The procedure of Example 2 was repeated, except that 0.33 part of 8-hydroxyquinoline was dissolved with 0.08 part of sodium hydroxide and added to the pulper, followed with 0.29 part of copper sulfate pentahydrate to precipitate 0.40 part of copper 8-quinolinolate. The sheet formed from this furnish contained from 0.26 to 0.37 part of copper 8-quinolinolate for every 100 parts of fibers. This sheet had essentially the same light transmittance and passed the same fungus resistance and soil burial tests as the sheet made according to Example 2.

The paper was coated on the smooth side of the crepe with a composition comprising 21.3 parts of aluminum powder and 8.5 parts of acrylic polymer emulsified with 70.2 parts of water. This operation was performed with a conventional air knife coater and was followed by drying in a tunnel dryer. The resulting aluminum coated mulch paper retains the fungus resistant properties of the uncoated mulch paper and, in addition, is effective in repelling aphids.

Moldproof Paper Containing Di(Phenylmercuric)Ammonium Salts of Aliphatic Carboxylic Acids

H.C. Crandall and A.R. Bowers; U.S. Patent 3,560,332; February 2, 1971; assigned to Mosinee Paper Mills Company describes a moldproof paper produced by adding water-soluble di(phenylmercuric)ammonium salts of aliphatic carboxylic acids to a partially or fully dried web of paper. Compounds useful in the process consist of water-soluble di(phenylmercuric)ammonium salts of aliphatic carboxylic acids. The term "water-soluble" as used is defined to mean that the compounds have a water solubility at 25°C of at least about 2.5 grams per 100 grams of solution. Specific compounds within this class which are effective are di(phenylmercuric)ammonium acetate, di(phenylmercuric)ammonium propionate, di(phenylmercuric)ammonium butyrate, and di(phenylmercuric)ammonium methacrylate. The di(phenylmercuric)ammonium propionate is particularly effective as to imparting moldproofness and has a relatively high degree of water-solubility (6.1 grams per 100 grams solution at 25°C); hence, it is considered the most useful compound within the above generic class for commercial practice.

In order to produce the treated papers, water solutions of the class of active compounds described above are to be applied to either a partially dried or fully dried web of paper on a papermaking machine.

The typical papermaking machine will include a sheet-forming section in which a web of paper is formed on a moving wire in the case of a Fourdrinier machine or on a cylinder mold with a cylinder machine, a press section which receives the web from the sheet-forming section and partially dries it by mechanical pressing or squeezing action, and a dryer section which removes additional water by the application of heat and which will include calenders by which the dried web is compacted and given a smooth finish.

The mildewproofing agents are initially water-soluble and it is therefore impractical to incorporate them in the stock or slurry which is fed to the papermaking machine prior to the formation of the paper web. Hence, it is essential that the materials be applied to a partially dried web of paper after it has advanced into the press section of a papermaking machine.

This is best accomplished by passing the partially dried web of paper through either a vertical or horizontal size press in order to apply the water solutions; for maximum pickup of solution, the web should be dried so that its moisture content is from 10 to 20%. The usual size press apparatus consists of rubber covered rolls positioned to be driven in contact with each other and adapted to be hydraulically loaded to increase or decrease the pressure at the nip along which the rolls are in contact.

In applying the solutions to a paper web having a width of approximately 100 inches, a pressure on the nip between the size press rolls of approximately 200 pounds per linear inch was found satisfactory, but this may be increased or decreased in order to change the characteristics of the sheet or the pickup of treating solution; thus, this pressure may be from about 50 pounds per linear inch up to 350 pounds per linear inch, or even higher.

When applied to a partially dried paper web, the solution pickup, calculated as a percentage of the dry weight of the paper being treated, can be from 25 to 200%, depending on the type of paper being treated, and within this range, from 60 to 80% is considered optimal.

Another possible point of application of the mildewproofing agents in order to form the treated papers is at the water box on the calender stack. In this instance, the sheet is dry at the point of application and the percent wet pickup is relatively low, so that the solution concentration required is relatively high. For example, in the case of a calender stack application of di(phenylmercuric)-ammonium propionate, when the wet pickup of the paper was 7%, a solution containing 0.11% of the compound was used for treating the paper which, after treatment included 0.008% of the active compound. After the specified compounds are applied to the paper, the web is dried by the application of heat to render the compounds at least substantially insoluble. The dried papers are resistant to leaching of the active compound.

After application of the specified solutions, the paper web is dried and the compounds become at least partially insoluble and may undergo chemical change. The temperature range to obtain this effect expressed in terms of the temperature of the dryer surfaces in the dryer section of the papermaking machine, is from 150° to 390°F, with 180° to 285°F being considered the optimum portion of this range.

The operative temperature range based on the actual temperature of the paper web is from 35° to 250°F, with 100° to 225°F considered optimum. Thus, the finished papers contain the material formed from heating the paper web to which the specified water-soluble salts have been applied at the preceding temperature ranges, and such materials may be insoluble heat conversion products of the salts or may be unconverted forms of the salts, or a mixture of both.

Examples 1 through 11: Comparison of Papers Treated with SPC and DPMAP – The table below shows the results of tests of treated papers with respect to growth of *Aspergillus terreus,* color of paper and odor.

					Mold Growth, % Coverage		
Example	Chemical	Quantity, %	Color of Paper	Odor of Paper	After 4 Days	After 7 Days	After 14 Days
1	SPC	0.5	Slightly pink	Slightly sharp	None	25	75
2	SPC	0.9	Pink	Sharp	None	10	30
3	SPC and whitener	0.9	Slightly pink	Sharp	None	10	30
4	SPC	1.2	Pink	Sharp	None	5	10
5	SPC	1.59	Pink	Sharp	None	None	5
6	DPMAP	0.005	White	No odor	None	None	5
7	DPMAP	0.008	White	No odor	None	None	None
8	DPMAP	0.016	White	No odor	None	None	None
9	DPMAP	0.030	White	No odor	None	None	None
10	DPMAP	0.048	White	No odor	None	None	None
11	DPMAP	0.09	White	No odor	None	None	None

Note: SPC = Sodium pentachlorophenate.
DPMAP = Di(phenylmercuric)ammonium propionate.

Example 12: Moldproof Starch Based Coating – In a jacketed kettle equipped with an agitator, a starch dispersion was prepared by adding 160 parts of oxidized starch to 480 parts of water. The mixture was stirred until the starch was well wet and suspended in the water after which the temperature was raised to 200°F and the starch cooked for 20 minutes. The temperature was lowered to 150°F.

In a kneading machine, a clay slip was prepared by adding, with mixing, 950 parts clay, 50 parts titanium dioxide, 1 part soda ash and 4 parts sodium metaphosphate to 293 parts of water. Mixing was continued until the slip was substantially free of lumps of solid materials. The starch dispersion, prepared as described, was added slowly with mixing to the clay slip along with 1 part sulfonated castor oil and 42 parts of a 10% soap solution. To the above color was added 0.048 part by weight of DPMAP. The coating was diluted with 1,800 parts of water and screened for application to paper. For on-machine coating, color prepared in this manner can be used, after screening, with or without dilution and for off-machine coating, it is preferably diluted and screened prior to application.

Water Repellent Fungicide of Wax, Reactive Polymer, Fatty Acid Ester, o-Phenylphenol, Amine Fatty Acid Soap

The process of *L.I. Winebrenner and J.S. Knauss; U.S. Patent 3,509,083; Apr. 28, 1970; assigned to U.S. Plywood-Champion Papers Inc.* relates to a protective solution to protect cellulose products, both the natural and manufactured products, from the destructive and degrading forces of water and biological organisms.

By cellulose products, both natural and manufactured, is meant among other materials, dimension wood, lumber, plywood, manufactured hardboard, timber, paper and especially paper used in the manufacture of construction materials, corrugated board, compressed cellulose fiberboard, and other cellulose products exposed to weather and biological destruction.

By the protection against destruction and degrading of these products is meant protection against warping, splitting and cracking of lumber and the like, by the forces of water; the degrading of lumber and the like by biological staining, mold, mildew and dirt; the swelling, curling and buckling of corrugated board, paper and compressed cellulose fiberboard, by the forces of water. These products are also protected against the loss of tensile strength due to the forces of water penetration and biological invasion.

The protective solution comprises a water based emulsion of a reactive polymer, fatty acid esters, waxes, a biocide, surface active agents and additives, which may be either easily diluted with water or used at full strength and applied or incorporated in the porous bodies. The degree of dilution with water is directly proportional to the degree of protection and may hence be controlled by the user to accomplish the degree of protection desired.

The emulsion concentrate, or a water dilution of the emulsion concentrate, may be applied directly to the surface of the lumber or other cellulose construction material by spray, roller coat, brush or dip application. The emulsion penetrates effectively into the wood or cellulose surface and becomes incorporated in and associated with the surface fibers of the matrix.

The solvent (water) will be lost by rapid evaporation, which results in an immediate breakdown of the emulsion. The organic solids phase of the emulsion remains incorporated in the surface fibers of the cellulose. Then follows a breakdown of the surface active agent; the amine component being either fugitive and lost by volatilization or reacting chemically with the cellulose.

This results in a layer of the very desirable remaining organic solids within the surface of the wood, which is entirely hydrophobic. Thus, a protective film is provided in a relatively short period of time, depending upon the temperature and relative humidity of the environment. There is a subsequent curing of the protective layer of the reactive polymer, greatly increasing the imperviousness of the layer to water. The biocidal agent in the emulsion remains in the surface of the cellulose product. A particularly desirable mixture of ingredients to be emulsified is as follows.

Ingredients	Range (weight percent)
Paraffin wax	4 - 7
Petrolatum wax	3 - 6
Reactive polymer	4 - 7
Terpene solvent	1 - 3
o-Phenylphenol	1 - 3
Fatty acid	3 - 5
Amine	1 - 5
Water dispersible lecithin	0.5 - 1
Pentaerythritol-fatty acid ester	2 - 4
Urea	2 - 5
Water	60 - 75

The protective concentrate is an oil-in-water emulsion, wherein the amount of water present in the concentrate may vary from about 60 to 70% and higher. A desirable specific concentrate composition, for example, is one which has about 64% water present, such as the following.

Ingredients	Weight Percent
Paraffin wax	6.29
Petrolatum wax	4.71
Reactive polymer	6.15
Pine oil	1.51
o-Phenylphenol	2.05
Fatty acid	4.27
Ammonium hydroxide 28%	4.38
Water dispersible lecithin	0.62
Pentaerythritol-fatty acid ester	3.17
Urea	3.08
Water	63.77
Total	100.00

The characteristics of the materials used in the above compositions are as follows.

Paraffin wax—120°/150°F AMP (American Melting Point), white crude scale wax, having a flash point of 410°F minimum, a penetration of 40/50, a viscosity at 210°F of 39/40, and oil content (ASTM) of 1.5 maximum, and a color (Saybolt) of +25.

Petrolatum wax—a technical amorphous petrolatum with a Saybolt melting point of 120°/140°F, an ASTM consistency of 150 to 180, a Saybolt viscosity at 210°F of 105, and a dark brownish color.

Reactive polymer—a reactive polymer of diolefins and olefins, a viscosity (SSU/210°F) of 100 to 250, an iodine number of 240, a flash point (Cleveland open cup) of 210°F minimum, and an ash content of 0.2%. This material available commercially as CTLA polymer (Enjay Chemical Co.).

Pine oil—a special synthetic pine oil, having a refractive index at 20°F of 1.480, a color Saybolt of +17, a flash point (Cleveland open cup) of 170°F, total alcohols of 73% and an ASTM distillation range of 197° to 222°C.

o-Phenylphenol—a biocide having the chemical formula $C_6H_4(C_6H_5)OH$, with a molecular weight of 170.2, freezing point of 57°C, boiling point of 286°C, flash point 255°F, fire point of 300°F.

Fatty acid—preferred are the C_9 to C_{19} acids (lauric, oleic, palmitic, stearic acids, etc.).

Ammonium hydroxide 28%—an ammonium hydroxide solution containing 29.4 to 29.8% NH_3 with gravity of 26.0° Bé.

Water dispersible lecithin—lecithin having 49% acetone insolubles, an acid value of 22 to 31, and a color maximum of 12 Gardner. A typical product is Kelecin 1080.

Pentaerythritol-fatty acid ester—a reaction product between a fatty acid and pentaerythritol alcohol, having an acid number of 25 maximum, of Gardner color of 100 maximum, and a Gardner-Holt viscosity of G to K.

The concentrate is very effective in the treatment of other fibrous materials such as gypsum board or plaster board or paper facing for this type board to render the same water repellent and water retardant. When the concentrate, as described above, is mixed with about 1 to 6 or more parts of water, and when used in the treatment of paper, the resistance of the paper toward water absorption is raised to a high degree, without interfering with the drying and adhesion of paint which may later be applied to the paper.

1,1-Diallylguanidine Salts for Inhibition of Mold, Bacteria and Algae

F.C. Schaefer and A.C. Wright; U.S. Patent 3,734,939; May 22, 1973; assigned to American Cyanamid Company found that 1,1-diallylguanidine, having the theoretical formula:

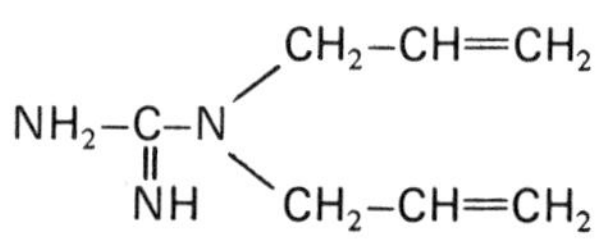

and salts thereof are useful as antimicrobial agents, as emulsifying agents, as detergents, as sizing agents for paper; and as raw materials for the manufacture of water-soluble cationic thermosetting strength resins for paper.

At room temperature 1,1-diallylguanidine free base is soluble in water, methanol, ethanol, butanol, acetonitrile and acetone. The acetate, chloroacetate, nitrate, hydrobromide, hydrochloride, maleamate and benzenesulfonate salts of 1,1-diallylguanidine are very water-soluble. Dilute solutions of the free base in any of the solvents named, and dilute solutions of the aforementioned salts of 1,1-diallylguanidine in water, when sprayed on sawdust, woodchips, paper, leather, and painted surfaces, inhibit the growth of algae, fungi and bacteria thereon. When used as a surface biocide, concentrations in the range of 0.1 to 0.5% are suitable for dry, nonabsorbent surfaces, and concentrations of 0.5 to 1% are suitable for moist and absorbent surfaces. The solvent evaporates and the residue inhibits microbial growth. 1,1-Diallylguanidine in free base and in dissolved salt form is cationic and is substantially adsorbed by cellulose fibers in dilute aqueous suspension.

Example 1: The Preparation of 1,1-Diallylguanidine Acetate – To 102 grams (1.05 mol) of diallylamine in an ice bath are added successively 60 grams (1.0 mol) of glacial acetic acid, 42 grams (1.0 mol) of anhydrous cyanamide and 50 milliliters of acetonitrile as solvent. The resulting solution is transferred to a dropping funnel and added at uniform rate over 90 minutes to 25 ml of refluxing acetonitrile.

After about a third of the solution has been added, product crystals appear and the mixture becomes progressively thicker. The temperature rises from 82° to 96°C during the addition. The final mixture is refluxed for 90 minutes, and cooled and filtered. The recovered white crystals are rinsed with acetonitrile and dried under vacuum, MP 214° to 216.5°C. The salt is soluble in water.

Example 2: The Manufacture of Paper Rendered Resistant to Microbial Growth by a Uniformly Adsorbed Content of 1,1-Diallylguanidine – An aqueous suspension of well-beaten cellulose papermaking fibers at a consistency of 0.6% is adjusted to pH 8, and to this is added 1% based on the dry weight of the fibers of 1,1-diallylguanidine (as the acetate) as a 1% solution in water. The suspension is briefly stirred, and the fibers are made into handsheets and dried according to standard laboratory procedure. The handsheets resist the growth of algae, molds, and bacteria thereon.

2-(C_8 to C_{18} Substantially Straight Chain)Pseudourea for Antimicrobial Activity

C.F. Hinz; U.S. Patent 3,728,213; April 17, 1973; assigned to American Cyanamid Company found that cellulose fibers which have an adsorbed content of a long chain alkyl pseudourea of borderline solubility such that it is substantially insoluble in water as the free base, but is soluble in acid possesses antimicrobial properties alone and in the presence of wet strength resins in the thermoset state.

These properties are retained when the paper is exposed to moist conditions for long periods of time at elevated temperatures such as are encountered in outdoor weathering. The paper also is sufficiently antimicrobial to strongly resist the growth of microbial life under incubating conditions, and the fibers carrying a pseudourea of the type described remain substantially sterile. Antibiotic bandages, diapers, bed sheets, boxes, surgeons' gowns, mulch paper for agricultural use, etc., can be prepared from paper composed of these fibers.

The microbial life against which the paper possesses activity includes the bacteria (including sulfate-reducing bacteria), yeasts, fungi and green algae, and lipophilic viruses. The pseduoureas referred to have the following formula:

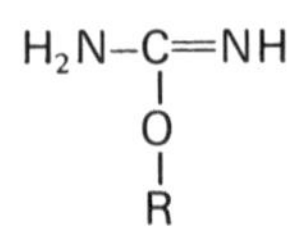

in which R represents a C_8 to C_{18} substantially straight chain alkyl substituent.

Satisfactory results are obtained when the fibers contain between about 0.3 and 0.5% by weight of the pseudourea as in this range the fibers (and the paper) possess very satisfactory antimicrobial activity yet over-use of the pseudourea is avoided. The optimum amount of pseudourea that need be present in any instance to procure the desired antimicrobial activity can be readily found by laboratory trial.

In many instances when the end product is paper or paperboard, it is desired that such product possess wet strenth properties. These properties can be imparted by adding any of the known cationic thermosetting wet strength resins for paper along with the pseudourea, or by applying the wet strength resin subsequently, for example, at the size press, the paper being subsequently heated to develop the wet strengthening properties of the resin.

Suitable resins include the melamine-formaldehyde resin colloid of U.S. Patent 2,345,543, and the polyamine-epichlorohydrin resins of U.S. Patent 2,595,935, and the polyamidepolyamine-epichlorohydrin resins of U.S. Patent 2,926,154.

It may be desired that the paper and paperboard possess superior dry strength without possessing permanent wet strength (to facilitate disposal of the paper after use). This can be accomplished by adding a water-soluble cationic or nonionic dry strengthening agent along with pseudourea. Suitable polymers for this purpose are disclosed in U.S. Patents 2,959,514 and 3,258,393.

Example 1: The Manufacture and Use of Mulch Paper – Into a furnish composed of a 50:50 mixture of hardwood and softwood papermaking fibers at pH 4.5 at a consistency of 0.6%, as it passes through the machine chest in a paper mill, is metered a 5% by weight solution of n-dodecylpseudourea hydrochloride at a sufficient rate to provide 0.5% of n-dodecylpseudourea (as the free base) based on the dry weight of the fibers and a solution of the melamine-formaldehyde wet strength acid colloid of U.S. Patent 2,345,543 at the same rate.

The furnish is processed into paper at a basis weight of 50 pounds per 25" x 40"/500 ream, and the paper is dried at 190° to 250°F for about one minute. Kjeldahl nitrogen analyses of the paper show that the dry paper contains 60 to 70% by weight of the dodecylpseudourea which had been added. From experience it is known that most of the remainder was adsorbed by the cellulose fibers, which were lost with the white water.

The resulting paper is transported to a pineapple plantation and is rolled upon the ground by an automatic pineapple seedling planter. Pineapple seedlings are then planted through approximately 1 inch diameter holes punched through the paper.

The paper largely prevents the growth of weeds around the seedlings, does not support microbiological growth during the period that the pineapples are growing, resists tearing when wet, and shortens the vegetation period of the pineapples. The paper loses virtually all its strength and disintegrates to fibrous state within three months after being ploughed under the ground.

Example 2: The Manufacture of an Infant's Diaper -- To an aqueous suspension of cellulose papermaking fibers at pH 4.5 and a consistency of 0.6% is added sufficient of an aqueous solution of 2-n-decylpseudourea acetate to provide 2%, based on the dry weight of the fibers. The fibers are formed into a web on a screen and the web is allowed to air dry. The web is carded to form cotton wool which is formed into a batt about 18" x 18" x ¼". The batt is placed between two sheets of wet strength paper 18" x 18", one of which is sized and the other is unsized. The edges of the two sheets are bound together with pressure-sensitive adhesive tape.

Example 3: The Manufacture of Highly Antimicrobial Paper of Excellent Wet Strength Properties – To an aqueous suspension of unbleached softwood kraft fibers at 0.6% consistency and pH 6 in ordinary (nonaseptic) laboratory water is added with gentle stirring sufficient of a 5% by weight solution of 2-n-dodecylpseudourea hydrochloride to supply ¾% of this compound based on the dry weight of the fibers.

There is then added sufficient of a 1% solution of a water-soluble cationic thermosetting glyoxylated 95:5 molar ratio acrylamide:diallyldimethylammonium chloride copolymer to supply 1% of the polymer based on the dry weight of the fibers. The suspension is then formed into handsheets at 50 pound basis weight which are dried for one minute on a drying roll having a surface temperature of 220°F.

The resulting paper carries about 0.6 and 0.8% by weight of the respective agents, and has a TAPPI wet tensile of 9 pounds per inch. A 6" x 6" sheet of the paper is then inoculated with stock laboratory cultures so as to place on the sheet cultures of a bacterium and a fungus, and the sheet is allowed to air dry at 80°F for 30 minutes. The sheet is then rolled in a plastic sheet and allowed to incubate in the dark for 48 hours. The sheet is then incubated and is found to be sterile.

RESISTING DETERIORATION OF ACID AND AGE

Use of Gaseous Hexamethylenetetramine to Resist Acid Induced Deterioration

The recognition of the severe problem of deterioration of documents printed on paper that is or becomes acidic has prompted the development of several different processes for deacidifying such paper so as to halt or slow its deterioration.

J.E. Kusterer, Jr. and J.D. Hind; U.S. Patent 3,703,353; November 21, 1972; assigned to Council on Library Resources, Inc. discovered that paper which is susceptible to acid induced deterioration can be effectively, safely, and relatively inexpensively deacidified, and thereby made resistant to such deterioration, if it is impregnated with gaseous hexamethylenetetramine. A surprising discovery regarding this process is that the gaseous hexamethylenetetramine penetrates even tightly closed books extremely thoroughly and very rapidly.

Hexamethylenetetramine, also known as hexamine, formin, aminoform and urotropin and having the chemical formula $(CH_2)_6N_4$, is a colorless, substantially odorless, normally solid, organic alkaline compound that can be structurally represented by the formula shown on the following page.

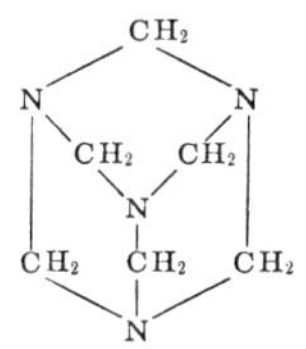

When the paper is sufficiently impregnated with the gaseous hexamethylenetetramine according to the process, its resistance to the deterioration that is induced by acidic conditions in paper is substantially increased. Preferably, the extent of impregnation will be sufficient to provide the paper with a pH (measured after the paper has been allowed to stand at room temperature and atmospheric pressure for 24 hours) of at least about 7.5, preferably at least about 8.3.

The pH values referred to are those of the resultant aqueous solution when one gram of one-eighth inch square pieces of the treated paper is soaked for one hour in 70 ml of distilled water at room temperature. As determined by this method, the initial pH value of paper which is susceptible to the deterioration that is alleviated by the process is usually about 5.0 or lower, and frequently about 4.3 or lower.

The temperature in the immediate vicinity of the paper, that is, the impregnation zone, during the treatment is sufficiently high, for example, at least about 35°C, that the hexamethylenetetramine can exist in the gaseous state long enough to diffuse through and/or condense on the paper. Preferably though, the impregnation zone temperature is not raised above about 42°C, most preferably not above about 40°C, because the higher the temperature to which paper is exposed, the greater it will be degraded by the heating.

One suitable manner of impregnating the paper with the gaseous hexamethylenetetramine is by reacting gaseous ammonia with gaseous formaldehyde while the reactants are in contact with the paper to be treated to yield the gaseous hexamethylenetetramine and H_2O, the reaction equation for which is as follows:

$$4NH_3 + 6CH_2O \longrightarrow (CH_2)_6N_4 + 6H_2O$$

In this manner the gas generation zone and paper impregnation zone are not separated. Alternatively, however, this reaction can be effected in a separate zone from, for example, an anteroom, to that zone in which the impregnation of the paper is to be effected, and the hexamethylenetetramine-containing product vapors then conducted into the impregnation zone.

The use of separate generation and impregnation zones has the advantage of permitting the use of lower temperatures in the impregnation zone due to cooling of the product vapors in the course of their transit from the generation zone to the paper to be treated in the impregnation zone.

The preferred temperature range for effecting the hexamethylenetetramine-yielding reaction is about 96° to 135°C. It is also preferred that the mixture of gases in the generation zone contain at least that amount of ammonia which is stoichiometrically required to convert all of the gaseous formaldehyde in the zone to hexamethylenetetramine. One purpose of this is to avoid as much as possible the situation in which formaldehyde polymerizes in, on, or near the paper being treated and leaves deposits of paraformaldehyde on or in the paper. Most preferably, then, the amount of gaseous ammonia in the generation zone will be maintained at about 101 to 103% of that which is stoichiometrically required to convert all of the gaseous formaldehyde in the zone to hexamethylenetetraamine.

A suitable source of gaseous formaldehyde which can be employed for producing the gaseous hexamethylenetetramine is that obtained by the decomposition of paraformaldehyde, which latter material decomposes to gaseous formaldehyde at about 90° to 135°C. This decomposition can be effected right in the generation zone, for example, by heating a slurry of paraformaldehyde in an inert, heat-transfer liquid such as mineral oil.

Advantageously, a decomposition catalyst such as phosphorous pentoxide is admixed with the paraformaldehyde in an amount of 1 to 5% (preferably 2 to 3%) based on the weight of the paraformaldehyde. Preferably, the mixture of paraformaldehyde and catalyst, when employed, is heated to about 115° to 125°C to effect release of gaseous formaldehyde.

When generating gaseous formaldehyde right in the generation zone by this method, it will usually be preferred to regulate the amount of paraformaldehyde used and the temperature at which it is heated so that it decomposes at the rate of about 0.015 to 0.018 milligram per minute per cubic centimeter of combined volume of the generation and impregnation zones.

It has been observed when treating certain types of bound books that unless the cover of the book is somehow shielded, for instance with paper toweling, a deposit can sometimes form thereon of white, powdery hexamethylenetetramine. The hexamethylenetetramine-yielding condensation reaction is facilitated if the reactant gases are agitated, for example, by a fan rotating at about 1,400 rpm. Use of a fan can also facilitate dispersal of the gaseous hexamethylenetetramine throughout the impregnation zone.

The process is described with reference to Figure 1.1, a single chamber apparatus. A book **16** to be deacidified is hung over rod **18** in chamber **13** and the access door **19** to the chamber is closed and secured. Chamber venting valve **17** is closed and a slurry **10** of paraformaldehyde in mineral oil is heated by electrically powered heating elements **11** while contained in pan **12** to release gaseous formaldehyde into the reaction-impregnation chamber **13**. Gaseous ammonia is simultaneously introduced to chamber **13** via line **14**.

The gases are thoroughly admixed by fan **15** while they react to form hexamethylenetetramine which impregnates the book. Condensation drip shield **20** prevents whatever water drops that form on chamber ceiling **21** from falling on the book. After conclusion of the treatment, heating of the paraformaldehyde-oil slurry **10** is ceased, as is the introduction of ammonia via line **14**, and unreacted gases are vented from the chamber by opening the chamber venting valve.

Example 1: A cloth bound book, the pages of which were composed of chemical-wood paper (50% hardwood and 50% softwood) having a pH of 4.2, was suspended in the impregnation zone of the deacidification apparatus depicted in Figure 1.1. The volume of the chamber of the apparatus was 22.157 liters.

FIGURE 1.1: SINGLE CHAMBER APPARATUS FOR GASEOUS DIFFUSION PAPER DEACIDIFICATION

a.

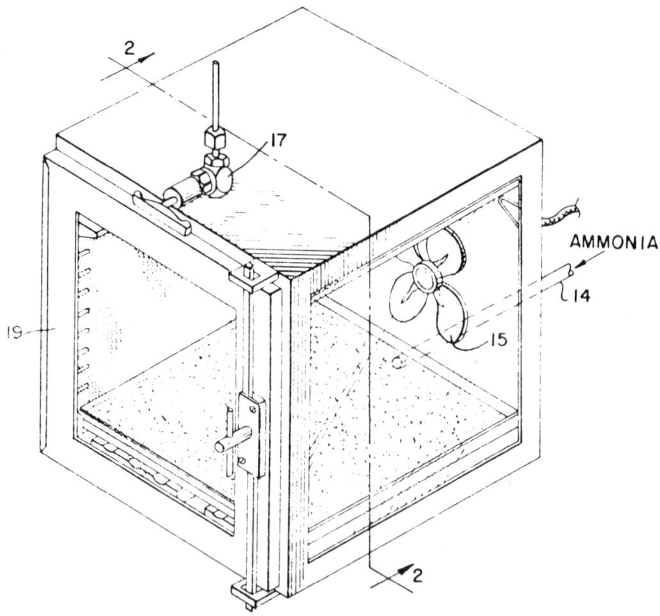

Isometric View of Single Chamber Apparatus

b.

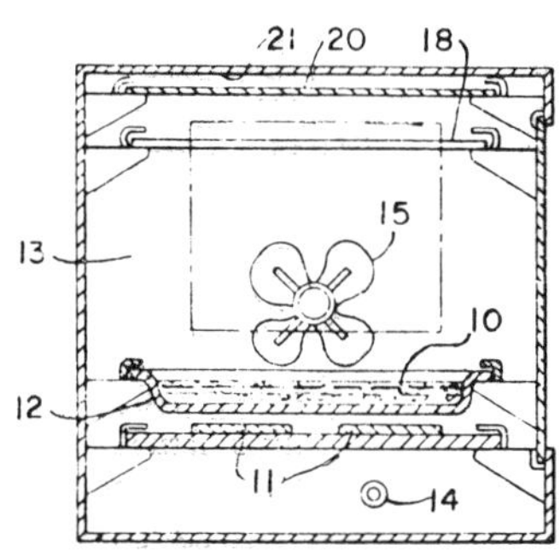

Front Sectional View Along Line **2-2** of (a)

(continued)

FIGURE 1.1: (continued)

c.

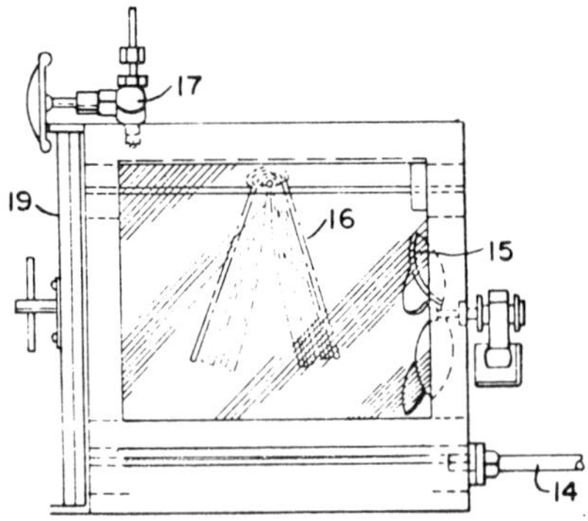

Side View of (a)

Source: U.S. Patent 3,703,353

Ammonia gas was passed through the chamber for 30 minutes at a flow rate of 0.12 mol per minute, while 40 grams of paraformaldehyde was slurried in a 135°C mineral oil bath in the bottom of the chamber. The chamber temperature in the vicinity of the book reaches about 42°C.

After two hours in the chamber, the paraformaldehyde having been substantially completely decomposed as evidenced by no appearance of white solid in the mineral oil bath, the book was removed and observed to have a strong smell of ammonia, most, if not all, of which left the book after it was allowed to sit open in a room overnight.

The pH of the pages of the book was then determined to be 8.3. The pages of the book were observed to be no more discolored than they were before the treatment. Portions of the book had a very slight amount of an oily film on them. The pH of the pages was measured again 4 days later, during which period the book was only subjected to ambient conditions, and was found to be 8.2, indicating the lasting effect of the deacidification process.

Example 2: Pages of another copy of the book used in Example 1, both before and after being subjected to a 2½ hour deacidification treatment, were subjected to fold endurance testing to determine the effect of the treatment on the strength of the paper.

Testing was with The Massachusetts Institute of Technology folding endurance tester which repeatedly bends a 15 mm wide strip of the paper to be tested across a line in a 270° arc, under one-half kilogram tension, until failure. The endurance was recorded as the average number of blends endured per strip, after testing 50 strips. Both machine directional (MD) fold endurance and cross directional (CD) fold endurance were determined in this manner. The results were that the deacidification treatment caused an average 26.2% reduction in the

CD fold endurance and 44.2% reduction in the MD fold endurance of the paper. The treated (deacidified) and untreated paper were then both subjected to an accelerated aging test, consisting of maintaining the paper at 68°C for 30 days, and the fold endurances of the thus aged paper determined. The fold endurance of the treated paper was substantially unaffected by the accelerated aging, exhibiting 102% of the CD fold endurance it possessed before aging, and 96% of the MD fold endurance it had before aging.

The results with the untreated paper were quite different, however. Aging of those samples caused a drop in their CD fold endurance to 64% of the pre-aging value, and a drop in their MD fold endurance to 61% of the pre-aging value.

These test data demonstrate that the deacidification treatment, while effecting an initial strength loss for the paper, provides resistance to further loss of strength and, overall, a much longer useful life than that to be expected for the untreated paper. Specifically, under the abovementioned aging conditions the treated paper had a half-life of greater than 150 days for both MD and CD fold endurance, while the untreated paper had only a 49 day half-life for CD fold endurance and a 42.5 day half-life for the MD fold endurance.

Deacidifying and Buffering Using Aqueous $CaCl_2$ and NH_4CO_3

J.C. Williams, G.B. Kelly, Jr. and R.L. Best; U.S. Patent 3,898,356; August 5, 1975; assigned to the U.S. Library of Congress describe a process of neutralizing the acidity of paper and buffering it on the alkaline side to improve its permanence. The method contemplates in serial order contacting paper with a solution of an aqueous soluble salt of a Group II metal (preferably water-soluble alkaline earth) and a salt capable of forming a relatively aqueous insoluble buffering compound in situ with the Group II metal.

The salts are preferably the water-soluble halides (preferably chloride) and acetates. When necessary the buffering agent is set into the paper by drying the paper prior to washing as in the case of partially soluble buffer salt. The paper is then washed in a controlled manner to remove excess reactants and the soluble reaction products and finally dried as a neutralized and buffered paper.

While either of the two solutions used may contact the paper first, it is preferred to use the suitable Group II metal salt solution first in most cases to allow the metal ion to react with cellulosic carboxyl groups which may be present. The reactant salt (preferably a carbonate) is an aqueous soluble material capable of interacting with the Group II metal salt to form a water-insoluble basic salt.

Washing is an important element of this procedure. The wash step should be of such a nature so that it does not dissolve deposited buffering material while physically removing gross surface deposits of excess materials and the soluble reaction products.

The wash water must be alkaline or it will dissolve the deposited alkaline earth compound. It is preferred that the wash water be saturated with the deposited alkaline earth compound. For example, when the deposited compound is calcium carbonate, the wash water is first flowed through a bed of calcium carbonate, such as marble chips or like material to saturate the water. A final rinse with distilled water is used before testing the effluent from the paper.

In general two solutions are prepared. Most preferably, the Group II metal salt solution is 10% calcium chloride and the reactant solution is 10% ammonium carbonate. The sheet of paper to be deacidified and buffered is dipped in the Group II metal salt solution and then in the reactant salt solution. Any number of methods of contacting the paper and solution are available. Besides dipping the paper in the solution, the solution may be sprayed or sponged onto either the obverse or reverse side of the paper.

One solution can be applied to one surface and the other solution applied to the reverse side of the paper causing the interaction to take place deep within the fiber structure of the paper. This produces the insoluble buffer such as calcium carbonate in situ in a desired amount from 0.3 to 10% dependent upon the concentration of the starting solutions and absorptive ability of the paper. The reaction also produces ammonium chloride which is subsequently removed. The treated sheet is next washed in running water until the wash water tests at a normal chloride content for the wash water.

Example: Three pounds of prepared rag and two pounds of bleached northern kraft were beaten to a Williams freeness of eight minutes. One percent Hercules Neuphor, a modified rosin and 3% dissolved papermakers' alum were added, based on 100 dry fiber. The stock was formed into handsheets, the slurry receiving further alum solution as required to give the dry sheet a pH of from 4.6 to 4.8.

The basis weight of the sheets was 75 grams per square meter, the brightness was 64 (according to TAPPI 452-OS-58, Standards and Suggested Methods, Technical Association of the Pulp and Paper Industry) and titration of 2.5 grams of paper opened in 250 grams of water showed 91 milliequivalents of acidity per kilogram. Such an acid paper shows a rapid drop in MIT folding endurance (TAPPI T511 su-69) on accelerated aging in the oven.

Sheets to be deacidified and buffered were wet out by being immersed in an aqueous 50% methanol solution. The sheets were then soaked 15 minutes in 6% aqueous calcium chloride solution and drained. The sheets were next immersed for 15 minutes in 6% ammonium carbonate solution and drained.

As a washing procedure, the sheets were rinsed separately in a shallow pan three times with deionized water, draining between each wash. Next, the sheets were soaked together in fresh deionized water for 15 minutes. Finally, each sheet was rinsed separately in fresh water. Drippings from the sheets were tested with 0.3% silver nitrate solution and turbidity was found. Washing was therefore resumed.

The sheets were soaked another 15 minutes in deionized water with agitation and then each sheet was washed with running deionized water in a pan tilted at 45°. After 2 minutes of continuous washing, the sheets were drained. Drippings now showed a negligible content of chloride ion when tested with the silver nitrate test.

The sheets were air dried and prepared for oven aging along with the untreated controls. Dry oven aging was carried out according to TAPPI T453 ts-63 except that 100°C was used. Generally 72 hours at this temperature is considered to be equivalent to 25 years at ambient conditions. The paper was also aged in an oven held at 90°C and 50% relative humidity. The aged and unaged sheets were analyzed with the results shown in the table below. The treatments can be seen to be very effective.

Electron microscope pictures were taken of the paper. Calcium carbonate particles could not be seen at a magnification of 100X but were visible at 1,000X in the treated sheet as well distributed round particles. The washings with deionized water had evidently partially solubilized the crystals and removed the corners.

	pH	Titration meq/kg	% $CaCO_3$ (Alkaline Reserve)	Folding Endurance	Brightness
Start					
Control	4.8	96	-	1,072	77.1
Treated	9.0	-	4.75	1,110	77.1
3 Days Aging (25 years equivalent)					
Dry oven					
Control	5.0	56	-	38	73.8
Treated	8.9	-	3.21	980	74.7
Humid oven					
Control	4.8	86	-	33	68.0
Treated	9.2	-	6.48	787	74.2
6 Days aging (50 years equivalent)					
Dry oven					
Control	5.5	45	-	9.6	-
Treated	9.2	-	3.50	524	-
Humid oven					
Control	5.2	80	-	9.2	-
Treated	9.2	-	5.60	475	-
9 Days aging (75 years equivalent)					
Dry oven					
Control	5.4	67	-	1.8	-
Treated	9.0	-	2.66	434	-
Humid oven					
Control	5.3	111	-	0.6	-
Treated	-	-	-	459	-
14 Days aging (116 years equivalent)					
Dry oven					
Control	5.0	48	-	0.4	55.0
Treated	8.7	-	3.58	158	65.0
Humid oven					
Control	4.5	150	-	0	30.0
Treated	8.9	-	5.48	53.5	65.0

Chlorinated Hydrocarbon Solution of Polyvinyl Acetal plus Basic Agent

In U.S. Patent 3,698,925, there are described means for restoring degradable sheet materials by treating such materials with polyvinyl acetals. The treatment consists essentially in immersing the materials to be treated in a solution of the copolymers in an organic solvent. These copolymers consist in percentage by weight of about 15 to 25% of polyvinyl alcohol, 1 to 3% polyvinyl acetate and 70 to 82% of one or more polyvinyl acetals selected from acetals of formaldehyde, acetaldehyde and butyraldehyde. The compositions and methods provide for safe and permanent restoration of many sheet materials. However, these solutions do not provide adequate protection against deterioration due to acidic materials. It is well known that the most significant cause of deterioration of papers and other sheet materials is the presence of such acidic materials, either in the paper itself (for example, alum) or in the atmosphere (for example, carbon dioxide, sulfur dioxide, and nitric oxides).

The process developed by *C.B. Hayworth; U.S. Patent 3,778,401; December 11, 1973; assigned to World Patent Development Corp.* provides stable solutions containing both a chlorinated hydrocarbon solution of polyvinyl acetals of the composition described above and an aqueous solution of a basic agent.

Preferably, the chlorinated hydrocarbon solvent contains 2 carbon atoms. Examples of such solvents are ethylene chloride, ethylidene dichloride, 1,1,1-trichloroethane, 1,1,2-trichloroethylene, 1,1,2,2-tetrachloroethylene and 1,1,1,2-tetrachloroethylene; 1,1,2-trichloroethylene is preferred.

Stable monophasic solutions of a 1,1,2-trichloroethylene solution of polyvinyl acetals and an aqueous solution of a basic agent are obtained by dissolving the polyvinyl acetal in a solution of trichloroethylene and either isopropanol or methyl isobutyl ketone or a mixture of isopropanol and methyl isobutyl ketone, and adding thereto an alcoholic solution of either an aqueous solution of the basic agent or of a hydrated basic agent.

In practice from about 3 to 6 grams of the polyvinyl acetal is dissolved in 100 cc of a solution containing from 60 to 95 cc of trichloroethylene and from 5 to 40 cc of the isopropanol or methyl isobutyl ketone or mixture thereof. If desired, 1 cc of water may be added in place of either 1 cc of the isopropanol or the ketone. Other water-miscible solvents such as methanol, ethanol, propanol, acetone or methyl ethyl ketone may be used in place of the isopropanol or the methyl isobutyl ketone.

To 100 cc of the solution of the polyvinyl acetal, 2 cc of an alcoholic solution of the basic agent dissolved in water is added. The basic agent, which may be a hydroxide of sodium, potassium, strontium, barium or lithium; a carbonate of sodium or potassium; a bicarbonate of sodium, potassium or lithium; an acetate of sodium, potassium, lithium, barium or magnesium; or hydrates of these compounds, is first dissolved in water and the aqueous solution then dissolved in about 100 cc of a water-miscible alcohol.

Suitable alcohols include methanol, ethanol, propanol, isopropanol and mixtures thereof. Instead of first dissolving the basic agent in water, a hydrate of the basic agent may be added to the alcohol. From about 0.5 to 20 grams of the basic agent can be present in about 100 cc of the alcoholic solution.

Example 1: Deacidification to pH 6 – Solution A consists of 3.5 grams of polyvinyl acetal dissolved in a mixture of the following:

	Cubic Centimeters
1,1,2-Trichloroethylene	79
Isopropanol	15

(continued)

	Cubic Centimeters
Methyl isobutyl ketone	5
Water	1

(a) One gram NaOH was dissolved in 5 cc water, added to 10 cc methanol and then diluted with 90 cc isopropanol. Two cubic centimeters of this solution was added to 100 cc of solution A.

(b) Ten grams $Ba(OH)_2 \cdot 8H_2O$ was dissolved in 100 cc methanol. Two cubic centimeters of this solution was added to 100 cc of solution A.

(c) Ten grams $Mg(CH_3COO)_2 \cdot 4H_2O$ was dissolved in 10 cc of water and 90 cc isopropanol added. Two cubic centimeters of this solution was added to 100 cc of solution A.

Newspapers, or any other acid papers such as old books or documents, showing a pH in the range of 3 to 4 when dipped in the above solutions will show a pH, after drying, of at least 6 to 6.5. The pH of the treated pages can be established either by a universal indicator or by cold water extraction whereby samples of the treated pages are extracted in distilled water for a period of one hour, and the pH of the distilled water is determined.

Example 2: Deacidification to pH Above 6.5 – If it is desired to raise the pH above 6.5 when magnesium acetate is used as the deacidification agent, the addition of small amounts of sodium hydroxide will be found effective.

Ten grams NaOH was dissolved in a mixture of 7.5 cc of water and 117.5 cc of methanol. One hundred and twenty-five cubic centimeters of isopropyl alcohol was then added. This is solution B.

Fifty grams of $Mg(CH_3COO)_2 \cdot 4H_2O$ was dissolved in 15 cc of water and 485 cc of ethyl alcohol. Then 500 cc of isopropyl alcohol was added, and finally 500 cubic centimeters of methyl alcohol. This is solution C.

To solution C, 150 cc of solution B was added with stirring and the mixture slightly warmed to dissolve precipitate which formed on addition. This is solution D.

The following polyvinyl acetal solution was prepared. Thirty-five grams of polyvinyl acetal was dissolved in a solution of 120 cc of isopropyl alcohol and 680 cubic centimeters of trichloroethylene. This is solution E.

Addition of various amounts of solution D to the polyvinyl acetal solution results in a clear solution that will neutralize and strengthen old, brittle and acidic papers. The effect on the pH of acid papers such as, for example, newspapers of various concentrations of solution D in solution E is shown in the table below. pH was measured by indicator after newspapers were dipped into solution for five minutes and dried.

- - - - - Cubic Centimeters of - - - - - -		
Solution E	**Solution D**	**pH**
100	0	4.0 - 4.5

(continued)

Cubic Centimeters of		
Solution E	Solution D	pH
100	16	6.5 - 7.0
100	20	7.0 - 7.5
100	32	8.0 - 8.5

Decreasing Aging with Starch Sulfite-Bisulfite Surface Size

Paper tends to deteriorate by any of a number of means to provide a product of darkened color and/or reduced brightness. Many impurities or chemical additives used in paper manufacture tend to darken over a period of time probably due to oxidation effects involving the paper and the chemical additives.

These problems are particularly aggravated in papers manufactured from paper pulps containing large amounts of lignin. Such pulps are commonly referred to as mechanical (groundwood), chemimechanical and semichemical to distinguish them from the purely chemical pulps. Lignin deteriorates rapidly with natural aging. Heat and ultraviolet light tend to accelerate the natural aging process still further. The problems associated with oxidized metallic ion impurities is also aggravated in high-lignin papers since, in addition to the colored oxidation products, the metal impurities also react with lignin to produce colored complexes.

It is also characteristic of papers containing large amounts of unbleached sulfite or chemimechanical pulp prepared by the sulfite-type of cooking process to age more rapidly (with corresponding increased rate of deterioration) than paper manufactured from pure bleached chemical pulp.

R.A. Silberman; U.S. Patent 3,619,355; November 9, 1971; assigned to Georgia-Pacific Corporation found that impregnation of the surface of dry paper, that is, paper from the dry end of the papermaking process containing up to about 10% water, with a water-soluble sulfite or bisulfite, or a mixture of these, will greatly assist in maintaining paper brightness, and greatly retard the aging of the paper from any or all of the sources discussed above.

A convenient method for impregnating the surface of the paper with the sulfite and/or bisulfite is to incorporate the water-soluble salt in the starch surface-sizing compositions used to treat dry-end paper. The soluble sulfite or soluble bisulfite is applied to the paper at a dosage of from about 5 to 20 pounds per ton of paper.

Any pH which is consistent with the other characteristics of the paper being treated can be employed; a pH in the range of about 5 to 8, and preferably about 6 to 7 has been found quite suitable.

Fresh pulp cooking liquor of the sodium, potassium and/or ammonium sulfite-bisulfite type is suitable for use as the source of sulfite or bisulfite. Typical cooking liquors of this kind may have a pH of about 5.9 to 6 and from about 0.5 to 18% sulfite-bisulfite (expressed as SO_2). The relative amount of sulfite and bisulfite may range from predominantly sulfite to predominantly bisulfite. The soluble sulfite and/or bisulfite may be incorporated in an aqueous starch, or starch-clay, surface-sizing composition. Starch solutions may be prepared by slurrying starch in water and cooking the slurry at, for example, 190°F for about 20 minutes. These starch solutions contain about 3 to 15% by weight of dissolved solids. In general, the surface-sizing solution is applied in an amount of about 160 gallons per ton of paper treated. This is usually increased by the amount of sulfite solution added to the usual sizing solution.

The following tests and/or methods were employed for the evaluation of properties and characteristics of treated papers.

Brightness—determined according to TAPPI testing standard T 453 ts-63.

Bursting Strength—determined according to TAPPI testing standard (Mullen) T 403 m-53.

Tearing Resistance—determined according to TAPPI testing standard (Tear) T 414 ts-65.

Brightness Reversion—determined as the difference between the initial brightness of a test specimen and the faded brightness, i.e., the brightness after subjecting the specimen to a treatment which tends to fade the specimen.

Paper SO_2 Content—determined by extracting 5 grams of sample with 150 ml of distilled water, measuring pH, acidifying with sulfuric acid, addition of starch-potassium iodide indicator, and titrating with 0.00125 N KIO_3 to the final lasting color change. The number of cubic centimeters of KIO_3 solution required is an indication of the amount of SO_2 in the sample.

The effect of aging on the several characteristics of the paper can be indicated by accelerated aging of the paper with comparison of the properties before and after aging. In the case of brightness, these effects are referred to as fading and/or brightness reversion.

Oven, or heat aging is effected by the general accelerated method described in TAPPI tests T 453 ts-63 and RC-280, modified to employ an oven temperature maintained in the range 215° to 220°F with the specimens held in the oven for 6 hours. Light aging is effected by exposing the test specimens to a bank of six sun lamps for 6 hours while maintaining the ambient temperature below 100°F.

Gas aging is effected by hanging the test specimens in a stream of cooled combustion gases from a bunsen burner operating wide open. The gases are cooled to a temperature below 100°F. The "paper weight" refers to the weight of a ream made up of 24" x 36" sheets.

Example 1: Effect of Soluble Sulfite-Bisulfite Treatment on Aging — Starch solution was applied to white tablet paper of 37 pound weight and to white form paper of 46 pound weight. The papers were made up of mixtures of chemimechanical, bleached kraft and bleached sulfite pulp. Part of the paper was also treated with sodium sulfite-bisulfite (fresh cooking liquor) in accordance with the practices of this process. After treatment, some specimens were subjected to accelerated aging which, in these examples, was conducted for 24 hours at 130°C. A comparison of the physical properties of the paper before and after aging, with sulfite-bisulfite treatment and without, is provided in Table 1.

TABLE 1

Paper Weight, lb	Sulfite-Bisulfite Treated	Original Tear	Oven-Aged Tear	Original Mullen	Oven-Aged Mullen
37	No	42	23	20.8	12.6
37	No	40	22	21.0	12.0
37	No	40	22	22.1	12.7
37	Yes	40	38	21.5	19.3
37	Yes	42	38	21.9	18.9
46	No	48-52	32-36	24.0	15.1
46	No	56-58	36-38	23.0	12.3
46	No	56-58	32-34	26.3	14.6
46	No	52-58	30-36	29.2	13.9
46	Yes	54-56	54-54	28.1	25.0
46	Yes	56-58	50-56	29.5	27.1
46	Yes	56-56	52-54	26.0	23.2
46	Yes	50-54	50-52	24.5	21.7

Example 2: The Effect of Soluble Sulfite-Bisulfite Treatment on Brightness – The foregoing procedure was applied to the continuous processing of white form paper. After continuous operation was established with a standard starch surface-sizing solution, fresh sodium sulfite-bisulfite cooking liquor was mixed with the starch solution in a proportion of 11 gallons of cooking liquor per 160 gallons of starch solution. This corresponded to a sulfite-bisulfite addition to the paper in the range of about 11 to 12 pounds per ton.

The brightness and brightness reversion on accelerated aging were determined before and during sulfite-bisulfite treatment. The amount of sulfite-bisulfite picked up by the paper and contained in the starch was monitored regularly and reported as cubic centimeters of 0.00125 N KIO_3 under the heading SO_2. The results of these tests are provided in Table 2.

TABLE 2

Starch Characteristics				Loss in Brightness After Aging by			Paper Characteristics	
pH	Percent Solids	SO_2	Original Brightness	Light	Gas	Heat	pH	SO_2
6.0	8.2	0.2	74.8	4.9	1.3	3.0	5.7	8.2
			74.9	4.9	1.3	3.0	5.7	7.4
			75.2	5.1	1.5	3.0	5.8	7.5
6.1	9.2	179.6	76.2	4.3	1.3	2.8	6.5	225.4
			77.1	4.2	1.5	2.4	6.6	176.9
6.1	8.2	151.8	77.2	4.1	1.5	2.6	6.5	157.9
6.0	8.2	155.4	76.3	4.5	2.0	2.6	6.5	150.0

PROTECTION AGAINST FIRE

Phenolic Condensate/Phosphorus-Containing Esters

A.F. Shepard and B.F. Dannels; U.S. Patent 3,409,571; November 5, 1968; assigned to Hooker Chemical Corporation found that esters of a pentavalent member of the phosphorus family and a phenol-aldehyde or phenol-ketone condensate are useful for the treatment of normally combustible cellulosic materials to render them fire retardant. These esters of a member of the phosphorus family and a phenol-aldehyde or phenol-ketone condensate are characterized in that:

(1) A major proportion of the moiety of the member of the phosphorus family has the following formula

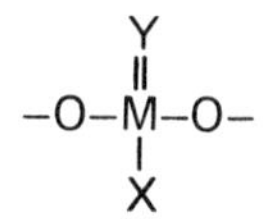

in which the unsatisfied bonds are attached to aryl nuclei of the same phenolic condensate, and in which M is an atom of the phosphorus family. Y is oxygen or sulfur, and X is halogen, hydroxyl, mercapto, hydrocarbyl, hydrocarbyloxy, halogen-substituted hydrocarbyl, halogen-substituted hydrocarbyloxy, or an aryloxy radical of the same phenolic condensate to which M is attached;

(2) At least 60% of the phenol-aldehyde or phenol-ketone condensate has o,o'-alkylidene linkages; and

(3) The phenolic condensate has an average number of aryl nuclei per molecule in the range of 2.2 to 8.

Suitable members of the phosphorus family of elements are arsenic, antimony, bismuth, and preferably phosphorus. Preferably, the phenolic condensates are novolacs, which contain more than one mol of the phenol per mol of the aldehyde or ketone.

Various reaction conditions can be employed for the reaction of the phenolic condensate and the compounds of the member of the phosphorus family, depending on the characteristics of the starting materials employed and the desired properties of the final products.

Generally, the temperature of the reaction is in the range of 100° to 250°C (preferably 150° to 200°C). Atmospheric pressure is usually employed for the reaction, but superatmospheric pressure or vacuum conditions can be employed, if desired. Reaction time can vary from 0.5 to 15 hours.

Various ratios of reactants can be employed, depending on the characteristics of the reactants and desired final products. Generally, up to about 0.5 mol of the compound of the phosphorus family is employed in the reaction mixture per equivalent of phenolic nucleus in the phenolic condensate. The ratio of reactants is preferably within the range of about 0.05 to 0.33 mol of the compound of the phosphorus family per equivalent of phenolic nucleus.

The ester compositions of the process are useful for the treatment of normally combustible cellulosic materials to render them fire retardant. Paper is the preferred cellulosic material, and all types of paper, made by any of the well known paper production processes, are contemplated. The ester composition is generally employed in an amount to provide 30 to 100% by weight based on the weight of the cellulosic material.

The cellulosic material to be treated is contacted with a solution of the esterification product in a suitable solvent, such as ketone, or a halocarbon, such as carbon tetrachloride, chloroform, methylchloroform, dichloroethylene, trichloroethylene, ethyl dibromide, propylene dibromide, and the like. Generally the solvents have a boiling point of less than 150°C.

The contacting step may be carried out in a variety of ways. The cellulosic material can be immersed in a tank containing the solution for a suitable period of time in a batchwise manner, or can be continuously passed through such a tank by means of rollers which facilitate the passage of a cellulosic sheet such as paper.

The composition can also be applied to cellulosic material by spraying, or by passing a sheet of the material through rollers that have been wetted with the solution. The phosphorus composition can be added to the beater in a papermaking process. The temperature of the process is preferably at room temperature, or about 30°C.

When the cellulosic material has been treated with the solution, the excess solution is drained or squeezed out, and the treated cellulosic material is dried at a temperature up to 150°C. Generally, a suitable curing agent, such as hexamethylenetetramine, is included in the treating solution. Then the dried, treated cellulosic material can be subjected to curing conditions to cure the ester composition.

Example 1: One hundred and thirty-six parts of high ortho content phenol-formaldehyde novolac (about 95% of o,o'-alkylidene linkage) having an average molecular weight of about 550 were placed in a reactor under a dry nitrogen atmosphere and heated to 170°C. One hundred parts of 2,4-dichlorophenylphosphorodichloridate were then added portionwise, while the temperature was slowly increased to 207°C.

After the HCl evolution stopped, the resin was poured into a receptacle and allowed to harden. The product was a hard resin that contained 5.3% phosphorus and 12% chlorine. When mixed with hexamethylenetetramine, the product readily cured at 165°C.

Paper was impregnated with a mixture of the foregoing resin and hexamethylenetetramine from a chloroform solution. After curing, the resin could not be washed out and the rate of burning of the paper was markedly reduced, as compared to a nontreated sample.

Example 2: Sixty-eight parts of the high ortho novolac used in Example 1 were heated to 180°C, with a slow stream of dry nitrogen passing over it. Fifty-two parts of p-bromophenylphosphorodichloridate were then added portionwise during a three hour period. The temperature was slowly increased to 211°C. The HCl evolved was trapped in a water scrubber, and 98% of the theoretical amount was found. The product was a hard, brittle resin which weighed 102 parts. It contained about 5.2% phosphorus and 13% bromine. When mixed with hexamethylenetetramine, the resin cured to a hard infusible material at about 165°C. The cured resin charred, but would not burn.

The resin was ground up and mixed with 10% hexamethylenetetramine. A 25% solution of this mixture in chloroform was prepared. Paper was then impregnated with this solution. After drying and curing, the paper burned only with difficulty.

Example 3: In the manner of Example 2, 47 parts of the high ortho novolac used were reacted with 40 parts of 2,4-dibromophenylphosphorodichloridate. The product was a clear amber brittle resin that contained about 24% bromine and 4.7% phosphorus. The cured resin charred when placed in a flame, but did not burst into flame. Paper treated with this resin was self-extinguishing.

Zinc Borate, Antimony Oxide, Organic Halide

S.B. Humphrey; U.S. Patent 3,770,577; November 6, 1973; assigned to Humphrey Chemical Corporation describes a composition which can be incorporated as a single ingredient into cellulosic materials such as paper or paperboard to provide superior fire retardant and afterglow retardant properties to the cellulosic material. The composition remains in the paper and cannot be leached out nor will it migrate.

The fire retardant composition comprises a mixture of about 2 to 75% by weight (preferably 15 to 50%) zinc borate, about 2 to 75 weight percent (preferably 15 to 50%) antimony oxide and about 15 to 85 weight percent (preferably 20 to 70%) of an organic halide source. The zinc borates employed can have the formulas $ZnO{\cdot}B_2O_3{\cdot}2H_2O$, $2ZnO{\cdot}3B_2O_3{\cdot}7H_2O$, and $3ZnO{\cdot}2B_2O_3{\cdot}5H_2O$. Generally these zinc borates will be in the form of a fine powder (maximum dimension of 100 mesh), and having a zinc to boron atomic ratio of from 1:1 to 1:4.

The antimony oxide employed in the composition can be either Sb_2O_3 or Sb_2O_4 and has a maximum particle size of 100 mesh (Tyler Screen). The minimum particle size of the zinc borate and antimony oxide is not critical. The organic halide source employed can be a suitable halogenated polymer or copolymer, for example, polyvinyl chloride (which is preferred) or polyvinylidene chloride or copolymers of vinyl chloride with vinyl acetate, vinylidene chloride, acrylates, etc.

The particle size of the organic halide source is of importance in imparting the desired fire retardancy to the paper or paperboard while retaining other desirable properties of the paper and minimizing the amount of fire retardant material needed. Generally for lighter weight papers the particle size of the organic halide should be small, for example, a polyvinyl chloride paste resin (PVC plastisol) having a typical particle size of less than 5 microns and preferably about 1 to 2 microns is most useful.

For paperboard products a larger particle size organic halide source can be employed for economic reasons. For example, polyvinyl chloride blending resin having a predominant particle size of about 25 to 40 microns but with some particles being even larger is useful. The amount of the fire retardant composition which is incorporated into the paper to provide fire retarding properties will be from about 5 to 50% of the total weight of the paper pulp and preferably about 15 to 30 weight percent. Generally, about 20 to 30 weight percent of the fire retardant composition is incorporated into lightweight papers, 10 weight percent of the composition is incorporated into heavier weight papers; and around 5 weight percent is often employed in very heavy weight products such as insulation board. In the case of insulation board, however, the bulk of the fire retardant composition is applied to the surface of the material.

The fire retardant composition is employed preferably as a dry powdered mixture and can be added to the furnish at any convenient point during the fabrication of the material from the beater to the headbox or cylinder vat as long as thorough dispersion is insured. Size press application and coating with suitable binders can also be employed. The pH is normally adjusted to about 4.5 to 5.0, for example, by addition of alum but this is not essential.

Generally about 0.1 to 0.5 weight percent based on the weight of the pulp of a retention aid is sufficient to insure the necessary retention of the fire retardant composition. The point of addition of the retention aid should generally be as close as possible to the headbox or cylinder vat in order to insure good mixing with a minimum of sheer or violent turbulence.

Example 1: Lab Scale Experiments – The quantities used were higher than would be required in commercial operation because hand sheets were employed.

No.	A	B	C	D	E	F	G	Results
1	30	20.0	15.4	-	-	50	-	Too brittle.
2	30	20.0	9.2	-	1.0	40	-	No flame, no glow.
3	30	12.0	12.2	-	1.0	40	-	Marginal flame, no glow.
4	30	-	9.2	-	-	40	5.0	Marginal flame, no glow.
5	30	-	-	-	13.4	25	-	Burns and glows.
6	30	-	-	6.0	2.0	40	4.0	No flame, no glow.
7	30	-	-	6.0	1.0	40	5.0	Marginal flame, no glow.
8	30	-	-	4.5	1.5	30	3.0	Marginal flame, no glow.

A = Wt dry pulp (grams).
B = 25% aqueous slurry of $ZnO{\cdot}B_2O_3{\cdot}2H_2O$ (grams).
C = Aqueous slurry (65% solids) of chlorinated paraffin resin containing 70% chlorine (grams) (Rezo Sperse).
D = Dry chlorinated paraffin resin containing 70% chlorine (grams) (Chlorez 700).
E = Antimony oxide.
F = Theoretical add on percent based on dry pulp.
G = Powdered $ZnO{\cdot}B_2O_3{\cdot}2H_2O$ (grams).

Example 2: Commercial Scale Experiments – (a) The following is the formulation of the fire retardant.

	Parts by Dry Weight
Zinc borate, $ZnO{\cdot}B_2O_3{\cdot}2H_2O$	16.67
Antimony trioxide	33.33
Polyvinyl chloride plastisol resin	50.00

(b) Using a single cylinder machine there was employed 2,000 pounds of pulp, 20 pounds of alum and 8 pounds of Nalco 632 (a retention aid). The fire retardant was added to the pulp after the alum while the Nalco 632 was just before the cylinder vat.

Six different samples of fire retardant were tried: (1) 600 pounds of the formulation of (a); (2) 500 pounds of the formulation of (a); and (3) 400 pounds of the formulation of (a). Samples (4), (5), and (6) were like samples (1), (2), and (3) except that a polyvinyl chloride blending resin (25 to 40 micron mean size) was employed in place of the polyvinyl chloride plastisol resin. In all six samples there were produced 0.030 inch sheets having a density of 1.0. The products were fire retardant in every case.

(c) Using a Fourdrinier machine there was employed 2,000 pounds of pulp, 25 pounds of alum, 7 pounds of Nalco 632 and 500 pounds of the formulation of (a). The fire retardant was added after the alum while the Nalco 632 was added just ahead of the headbox. Fifty pound basis weight sheet was produced which was fire retardant.

(d) Using a six-cylinder machine, there was employed 2,000 pounds of pulp, 9 pounds of Nalco 632, 500 pounds of the composition of (a) in which the plastisol, however, was replaced by the polyvinyl chloride blending resin (25 to 40 micron mean particle size). There was produced a fire retardant sheet having a thickness of 0.014 inch and a density of 1.0. The Nalco 632 was divided into six equal portions with one portion going into each of the six-cylinder vats.

Bonding Agent Containing Two Metal Bromides

M. Lewin; U.S. Patent 3,484,340; December 16, 1969; assigned to the Israel Minister of Trade and Industry found that inorganic bromides and oxybromides are very efficient fire resisting agents and that in distinction from other salts they can be incorporated in conventional aqueous bonding agent compositions in an amount sufficient to produce fire resistance without causing any deterioration of the bonding composition by coagulation of the bonding agent.

Every aqueous bonding agent composition has in fluid state certain optimal values for the viscosity and surface tension and any appreciable deviation from these values by way of increase or decrease, affects unfavorably the quality of the bonding agent composition and consequently also the quality of the bonded cellulosic article bonded therewith. Hereinafter, the viscosity and surface tension of a fluid aqueous bonding agent composition will be referred to for short as "fluid state characteristics."

The process provides, in the production of a bonded cellulosic article (paper, cardboard, veneer, chipboard) employing an aqueous bonding agent composition, a method of imparting a degree of fire resistance of the finished article by incorporating in the fluid composition at least two substances selected from the group of inorganic bromides and inorganic oxybromides provided that at least one of the substances is a bromide, one of which substances is capable of lowering at least one of the fluid state characteristics of the bonding agent while the other is capable of raising the same characteristic.

The substances are added in such a total quantity as to impart to the product a desired degree of fire resistance and in such a relative proportion that the fluid state characteristic assumes a selected and preferred value. In some cases the selected and preferred value of the reference fluid state characteristic may, for example, be substantially the value it had before the addition of the substances. In other cases the bonding capacity of the bonding agent composition may be improved by selecting a value that differs from the one before the addition of the substances. In any case the imparting of fire resistance is accompanied by a control of the viscosity and/or surface tension of the aqueous bonding agent composition and thereby a control of the bonding capacity.

It has been found that ammonium bromide lowers both the surface tension and the viscosity of an aqueous urea-formaldehyde resin composition. Against this it has been found that sodium bromide raises both the surface tension and the viscosity of the same composition.

The amount of oxybromides and/or bromides to be incorporated in the bonding agent will vary between 0.4 and 12% by weight of the cellulose or lignocellulose. Bromide salts that can be used are the bromides of sodium, potassium, ammonium, calcium, magnesium, zinc, cobalt, lead antimony and iron. Examples of oxybromides are those of calcium and magnesium.

Example: An adhesive mixture was made up of 30 parts of solid commercial starch and urea-formaldehyde containing adhesive (relative proportion approximately 1:1), 31.5 parts of ammonium oxide, 9.9 parts of sodium bromide and 70 parts of water.

Sixty grams of this composition were evenly brushed on a one square meter sheet of commercial kraft paper weighing 120 grams. A second similar sheet of kraft paper was placed on the first so that the adhesive line was between the sheets. The resulting laminate was pressed and dried for 3 minutes at 70°C. The analysis of a part of the laminate shows that it contained 6.5% of bromine.

A piece of the sample 2.75 inches wide and 8.25 inches long was tested according to the TAPPI method for flammability of treated paper and paperboard, T 461 mm-48. The sample did not inflame. When taken out from the Bunsen flame it did not continue to burn and no perceptible after-glowing was observed. A similar laminate composed of 2 one square meter sheets of kraft paper was prepared using the same adhesive to which no bromides were added. When testing the laminate according to the same T 461 mm-48, it inflamed immediately and burned completely.

Flame Resistant Paper Having High Delamination Resistance

A. Karimori, R. Kodama and A. Kato; U.S. Patent 3,834,937; September 10, 1974; assigned to Kabushiki Kaisha Kohjin, Japan describe a method of preparing flame-resistant paper having a high delamination resistance which comprises soaking paper in a dispersion containing a disperse phase of a synthetic rubber and/or a synthetic resin and, if desired, an alkyl ketene dimer, and a dispersion medium consisting of an aqueous solution containing a nonionic surface active agent as a dispersion stabilizer for the dispersed phase and a guanidine salt and/or a water-soluble salt of guanylurea as a flame retardant, thus impregnating the paper with the impregnating composition and incorporating the flame retardant into the paper in the amount of 10 to 50% based on the weight of the paper.

As the synthetic rubber to be used, butadiene-methylmethacrylate copolymers, styrene-butadiene copolymers, acrylonitrile-butadiene copolymers and chloroprene polymers are preferred, none of which comprise a compound having a polar group such as a carboxyl group.

Preferred synthetic resins are acrylic acid ester polymers, acrylic acid ester copolymers, copolymers of an acrylic acid ester with styrene, vinyl chloride polymers, vinyl chloride copolymers, vinyl acetate polymers, vinyl acetate copolymers, vinylidene chloride polymers and vinylidene chloride copolymers, each of which is emulsified with a nonionic emulsion stabilizer.

As the nonionic surface active agent, polyoxyethylene alkyl ethers and polyoxyethylene alkyl phenol ethers are preferred. As the guanidine salt, guanidine sulfamate and guanidine phosphate are preferably used. Guanylurea phosphate is preferably used as the water-soluble salt of guanylurea.

The dispersion to be used as the impregnating composition may be prepared by mixing a preliminarily emulsified synthetic rubber latex and/or a synthetic resin emulsion with an aqueous solution formed by dissolving the nonionic surface-active agent and the flame retardant in water. An alkyl ketene dimer may be added to the above dispersed solution so as to impart water resistance and sizing effect to the resultant flame resistant paper.

Further, in order to improve the water resistance of the flame resistant paper, a urea-formaldehyde resin or melamine-formaldehyde resin may be added to the above dispersed solution.

The composition (based on the solids) of the impregnating agent to be used is as follows:

	Parts by Weight
Flame retardant	20 - 40
Nonionic surface active agent	0.02 - 0.3
Synthetic rubber and/or synthetic resin	5 - 30
Alkyl ketene dimer	0 - 1.5
Melamine-formaldehyde resin or urea-formaldehyde resin	0 - 7
Catalyst for melamine-aldehyde resin or urea-formaldehyde resin, up to 10% based on the resin	

In the preparation of the impregnating agent, both the flame retardant and the nonionic surface active agent may be dissolved in warm water maintained at 80°C or lower, but when other components are dissolved simultaneously, it is preferable to maintain the liquid temperature below 30°C.

The effect of the alkyl ketene dimer sizing agent decreases especially rapidly with the passage of time when the liquid temperature is high. However, when guanylurea phosphate alone is used as the flame retardant, since its solubility is low, it is preferable to elevate the liquid temperature up to 50°C.

Examples 1 through 7: Base paper manufactured from a raw pulp composed of 80% kraft pulp and 20% sulfite pulp and a small amount of a melamine-formaldehyde resin as a wet strength-improving resin was soaked in an impregnating agent formed by dispersing the compounds indicated in the table in water, and squeezed so that the amount of solids to be fixed would be 45 to 50% based on the weight of the base paper. Results of the tests of physical properties of the flame resistant papers produced in this manner are shown in the table on the following page.

In each of the flame resistant papers obtained in the runs, improvement of the delamination resistance was observed. Especially, products of Examples 1, 2 and 4 treated with an impregnating agent comprising an alkyl ketene dimer were sized and a water resistance was imparted to them.

Impregnating Agent Composition (parts by weight based on solids)	Example Number						
	1	2	3	4	5	6	7
Guanidine sulfamate	35	35	35	35	30	35	35
Acrylic acid ester polymer	10	-	-	-	-	-	-
Acrylic acid ester/styrene copolymer	-	10	-	-	-	-	-
Chloroprene polymer	-	-	4	-	-	-	-
Butadiene/methylmethacrylate copolymer	-	-	7	-	-	5	5
Vinyl acetate polymer	-	-	-	5	-	-	-
Vinyl acetate/vinyl versatate copolymer	-	-	-	5	-	-	-
Vinylidene chloride polymer	-	-	-	-	3	-	-
Acrylic ester/vinylidene chloride copolymer	-	-	-	-	8	-	-
Butadiene/styrene copolymer	-	-	-	-	-	5	-
Butadiene/acrylonitrile copolymer	-	-	-	-	-	-	5
Polyoxyethylene nonyl phenol ether (nonionic surface active agent)	0.1	0.1	0.2	0.1	0.2	0.2	0.2
Alkyl ketene dimer	1.0	0.7	-	1.0	-	-	1.0
Melamine/formaldehyde resin	2.0	2.0	-	-	-	-	-
Organic amine salt catalyst for melamine/formaldehyde resin	0.2	0.2	-	-	-	-	-
Urea/formaldehyde resin	-	-	-	-	-	-	2.0
Organic amine salt catalyst for urea/formaldehyde resin	-	-	-	-	-	-	0.2

Physical Properties	Example Number							Base Paper
	1	2	3	4	5	6	7	
Weight (g/m^2)	116	116	117	117	119	118	117	80
Thickness (mm)	0.17	0.17	0.18	0.18	0.18	0.18	0.18	0.15
Tensile strength (machine direction) (kg/15 mm width)	6.0	6.4	5.8	6.2	5.9	6.8	6.5	-
Wet strength (kg/15 mm width)	1.5	1.5	1.1	1.2	1.3	1.3	1.2	-
Sizing degree (seconds)	6	12	-	8	-	-	-	-
Delamination resistance (machine direction) (kg/15 mm width)	200	210	180	205	170	210	208	84
Length carbonized (cm) (JIS A-1322 method)	7.1	7.4	7.3	7.5	7.2	7.5	7.5	-

DECORATIVE PAPERS

COLORED PAPERS

Montmorillonite Clay for Increasing Retention of Cationic Dyes on Cellulose Fibers

Perhaps the most used procedure for preparing colored paper is to add a dye to the fibers prior to their being formed into sheets. Such procedure is commonly referred to as beater dyeing. The open top and vigorous mixing action of the beater make it an excellent place to add the dyes. However, the term beater dyeing is taken to include also dyeing in Hydropulpers, Dynopulpers, Jordan's mixing chests, mixing tanks, fan pumps, head boxes and the like where similar mixing conditions exist.

The main classes of dyes used in the production of colored papers, and especially in beater dyeing, are the acid, direct and basic dyes. The first two classes, namely the acid and direct dyes, are anionic in nature and are generally sodium salts of color acids. Basic dyes are salts of color bases and in most cases are either the hydrochlorides or oxalates. They are cationic in nature and are readily soluble in water, especially under acid or neutral conditions.

While these basic dyes have strong tinctorial value and brilliancy, they are not always retained to the desired extent by the fibers and especially by bleached fibers. Additionally, many of such dyes have a tendency to give feathers or hairs when used to dye even unbleached fibers. This effect, sometimes also termed mottling, can be undesirable, especially in the pastel shades.

It is apparently caused by a small amount of the fibers coming into contact with an excessive concentration of dye or by certain fibers dyeing more readily than other fibers. The fibers absorb and retain such a quantity of dye that they are very heavily colored. Since there is little, if any, transfer or distribution of dye from the colored to the uncolored fibers, subsequent mixing of these heavily dyed fibers with the rest of the furnish does not result in a uniform distribution of the dye.

W.S. Baldwin and J.L. Keen; U.S. Patent 3,597,304; August 3, 1971; assigned to General Mills, Inc. found that the retention of cationic dyes by cellulosic fibers can be improved if the fibers are treated with a combination of the cationic dye and an alkali metal or acid montmorillonite clay.

The cationic dye and alkali metal or acid montmorillonite clay are preferably added to dilute dispersions of the cellulosic fibers prior to the formation of sheets from such dispersions. Preferably, the addition is made to the beater or refiner or to the already beaten or refined fiber.

The alkali metal (preferably sodium) or acid montmorillonite clay is used in an amount sufficient to increase the retention of the cationic dyes on the fibers. Preferably, such clays are used in an amount of about 0.05 to 1.0% by weight based on the dry weight of the fibers. Any of a wide variety of cationic dyes can be used. Preferably the dye is used in an amount of about 0.05 to 1.0% by weight based on the dry weight of the fibers.

Both the alkali metal or acid montmorillonite clay and the cationic dye are added to the fibers as dilute aqueous dispersions or solutions. Preferably, the dispersions or solutions contain less than about 10% by weight of the montmorillonite clay and/or dye.

When the materials are to be added to the fibers prior to the formation of sheets therefrom, they are also first preferably made up as dilute dispersions or solutions and then added to the aqueous fiber dispersion. In many instances, it is preferred that the montmorillonite clay be first added to the fibers followed by addition of the cationic dye.

Example 1: One-liter samples of an aqueous dispersion of moderately refined bleached kraft pulp at 0.8% solids adjusted to a pH of about 4.5 were first prepared. To one sample was added 2 cc of a 1% slurry of sodium montmorillonite clay (Wyoming bentonite) in water. Such slurry was prepared by dispersing 7% of the bentonite with water, passing the mixture through a centrifuge to remove grits and diluting with water to a 1% slurry.

To a second sample of the fiber dispersion was added 2 cc of a slurry of china clay (kaolin). To each of the above samples and a third sample containing no clay was added 4-cc portions of a 0.2% aqueous solution of the basic dye New Fuchsine. The dispersions were thoroughly mixed and then single handsheets were made from such pulp samples on a Noble and Wood handsheet machine at 0.025% pulp consistency.

The wet sheets were dried on a rotary dryer at between 200° to 250°F until they were dry, i.e., contained less than 10% moisture. The sheet containing the sodium montmorillonite clay was deeply and evenly dyed while the sheets containing no clay and the china clay were approximately the same color and very pale. They also showed feathering or mottling.

Example 2: Example 1 was repeated except that the pH of the fiber dispersion was maintained at about 8.5. Substantially the same results were obtained, with the sheet containing the sodium montmorillonite clay being far superior, i.e., more deeply and uniformly dyed with little or no mottling.

Example 3: Examples 1 and 2 were repeated except that the New Fuchsine dye was replaced by the following dyes: Green MX, Auramine OO, Chrysoidine Y, methyl violet, and methylene blue. The sheets containing the sodium montmorillonite clay all showed the same excellent dyeing as in Examples 1 and 2 with the exception that the Chrysoidine Y was apparently insoluble at pH 8.5 and thus the handsheet prepared with it was not dyed to any appreciable extent.

The sheets containing the sodium montmorillonite clay were all vastly superior to those containing the kaolin clay with the exception that the handsheets prepared using methylene blue at a pulp pH of 4.5 were approximately the same.

Fixing Anionic Dyestuffs with Cationic Polymeric Fatty Acid Polyalkylene Polyamines

The acid dyes which are anionic in nature have little, if any, affinity for cellulose. Accordingly, they have most commonly been employed where the furnish also contains alum or both alum and rosin. *J.L. Keen; U.S. Patent 3,619,356; November 9, 1971; assigned to General Mills, Inc.* discovered that the retention of anionic dyes by cellulosic fibers can be improved if the fibers are treated with a combination of the anionic dye and a polyamine derived from a polymeric fat acid.

The polymeric fat acid based polyamines are those having at least one polymeric fat acid radical and at least two primary amine groups. The preferred amines are those having the following formulas:

$$H_2N{-}CH_2{-}D{-}CH_2{-}NH_2$$

$$H_2N{-}(H_{2m}C_m){-}\underset{H}{N}{-}CH_2{-}D{-}CH_2{-}\underset{H}{N}{-}(C_mH_{2m}){-}NH_2$$

where m is an integer of 3 or 4, and D is the divalent hydrocarbon radical of a dimeric fat acid.

The polyamines derived from the polymeric fat acids are used in an amount sufficient to increase the retention of the anionic dyes on the fibers, preferably about 0.05 to 1.0% by weight based on the dry weight of the fibers. Any of a wide variety of anionic dyes can be used, preferably about 0.05 to 1.0% by weight based on the dry weight of the fibers.

The anionic dye is preferably added to the fiber in the form of a dilute aqueous solution or dispersion. Likewise, the polyamine is also preferably added in the form of a dilute aqueous solution or dispersion. Solubilization of the polyamine can be effected by at least partially neutralizing the same with a water-soluble organic or inorganic acid such as acetic acid and mineral acids including hydrochloric, sulfuric and the like.

The use of the dilute solutions or dispersions of dye and polyamine results in a more even distribution thereof on the fibers. It is also preferred that the solutions or dispersions contain less than about 10% by weight of the polyamine and/or dye. In many instances, it is preferred that the polyamine be first added to the fibers followed by addition of the anionic dye. In the preferred procedure wherein the polyamine derived from polymeric fat acids and the anionic dye are thoroughly mixed with an aqueous pulp or fiber dispersion, sheets can then be prepared using conventional techniques. Conventional additives such as fillers and the like can be added.

Example 1: One-liter samples of an aqueous dispersion of moderately refined, bleached kraft pulp at 0.8% solids at the pH of local tap water (about 7.9) were first prepared. To one sample was added 15 cc of a 0.2% aqueous solution of the acid dye Calcocid Blue 2G (a triphenylmethane dye, Color Index 42090). To a second sample was added 15 cc of the dye solution and 4 cc of a 1% aqueous solution of a diamine (1 cc diamine/100 cc water containing 0.2 cc acetic acid) of the first of the two formulas on the preceding page derived as indicated from the mixture of C_{36} acids prepared by polymerizing tall oil fatty acids.

The dispersions were thoroughly mixed and then single handsheets were made from such pulp samples on a Noble and Wood handsheet machine. The wet sheets were dried on a rotary dryer at between 200° to 250°F until they contained less than about 10% moisture. The sheets obtained from the dispersion containing the diamine were of a pretty pastel blue color whereas the sheets obtained from the dispersion containing only the dye were not dyed to any noticeable extent, being creamy white.

Example 2: Handsheets were prepared as in Example 1 except that the diamine was replaced by a polyamine of the second formula on the preceding page where m is 3 and D is the same as in the diamine of Example 1. The resulting handsheets had a relatively deep blue color indicating that even more of the dye was retained than with the diamine as used in Example 1.

Anionic Dye Used in Conjunction with Montmorillonite Clay and Polymerized Fatty Nitrogen Compound

In addition, *J.L. Keen; U.S. Patent 3,619,357; November 9, 1971; assigned to General Mills, Inc.* has also found that the retention of anionic dyes by cellulosic fibers can be improved if the fibers are treated with a combination of anionic dye, certain fatty nitrogen compounds and an alkali metal or acid montmorillonite clay.

The fatty nitrogen compounds useful in the process are fatty amines and quaternaries derived from fat acids and polymeric fat acids. The preferred fatty amines are those having the following formulas:

$$\text{(1)}\qquad R^2\text{—}\underset{\substack{|\\R^1}}{\overset{\substack{R^3\\|}}{N}}$$

$$\text{(2)}\qquad R^1\text{—}\overset{\substack{(R^2)_{2-p}\\|}}{N}\text{—}(C_mH_{2m}\text{—}\overset{\substack{H\\|}}{N}\text{—}H)_p$$

$$\text{(3)}\qquad R\text{—}[CH_2NH_2]_n$$

$$\text{(4)}\qquad R\text{—}\left[CH_2\text{—}\overset{\substack{(H)_{2-p}\\|}}{N}\text{—}(C_mH_{2m}\text{—}\overset{\substack{H\\|}}{N}\text{—}H)_p\right]_n$$

(5) 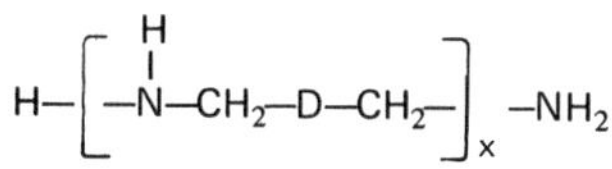

where p is an integer of 1 or 2, n is an integer of 2 to about 4, m is an integer of 3 or 4, x is an integer of 2 to about 40 or higher, R^1 is an aliphatic hydrocarbon radical of 8 to 22 carbon atoms, R^2 and R^3 are hydrogen or aliphatic hydrocarbon radicals or 8 to 22 carbon atoms, R is the polyvalent hydrocarbon radical of a polymeric fat acid and D is the divalent hydrocarbon radical of a dimerized fat acid.

The preferred fat acid based quaternaries are those having the following formulas

(6) 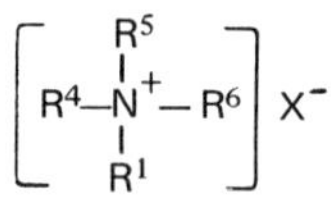

(7) 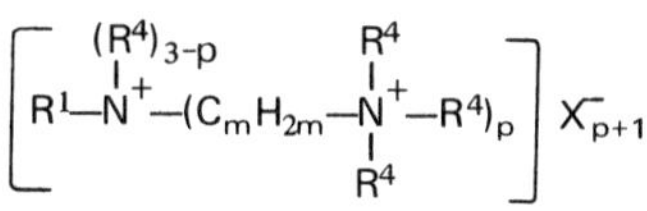

(8) 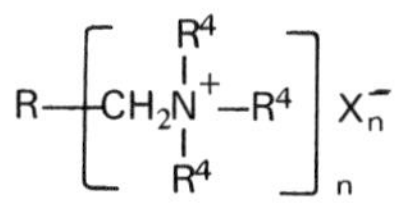

(9) 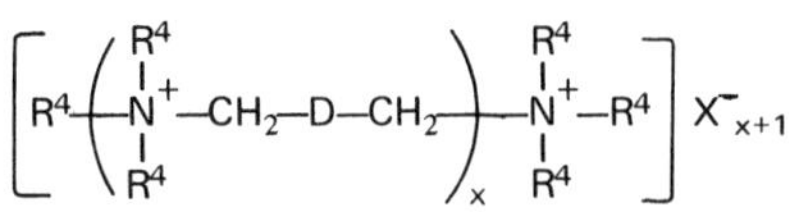

where p, n, m, x, R^1, R and D are as defined above, X^1 is a quaternary ammonium anion, R^4 is an alkyl group of one to six carbon atoms or the benzyl radical or an aliphatic hydrocarbon substituted benzyl radical wherein the aliphatic substituent contains from one to six carbon atoms and R^5 and R^6 are aliphatic hydrocarbon radicals of one to twenty-two carbon atoms.

Various of the amines of the formulas (1) and (2) are commercially available. The amines of the formulas (3), (4) and (5) are derived from polymeric fat acids. Such polymerized fat acids are prepared by polymerizing ethylenically unsaturated monobasic carboxylic acids having 16 to 22 carbon atoms or the lower alkyl esters. The polymerized fat acids are converted to the corresponding polynitriles by reaction with ammonia under nitrile-forming conditions.

The polyamines of the formula (3) above are prepared by hydrogenating the polynitriles in the presence of ammonia. Polyamines of this type are commercially available materials. Their preparation is further described in U.S. Patent 3,010,782.

Polyamines of the formula (4) are also commercially available. Such polyamines are prepared by reacting the compounds of formula (3) with an α,β-unsaturated nitrile such as acrylonitrile, crotononitrile, and methacrylonitrile and then catalytically reducing the resulting polynitrile. The preparation of these compounds is also described in the above-identified patent and in U.S. Patent 3,299,138.

The polyamines of the formula (5) are prepared by the condensation polymerization of the dinitriles under secondary-amine-forming conditions. Various of these polyamines are also commercially available and their preparation is further described in U.S. Patent 3,217,028.

The preparation of various quaternary ammonium compounds as discussed above is described in further detail in U.S. Patent 3,235,596 and in U.S. Patent 3,131,998.

Any of a wide variety of anionic dyes can be used. The dye is used in an amount of about 0.05 to 1.0% by weight based on the dry weight of the fibers. The fatty nitrogen compound and the alkali metal or acid montmorillonite clay are used in an amount sufficient to increase the retention of the anionic dyes by the fibers. The clay is used in an amount of about 0.05 to 1.0% by weight and the fatty nitrogen compound in an amount of about 0.05 to 1.0% by weight, such percentages being based on the dry weight of the fibers.

The anionic dye is preferably added to the fibers in the form of a dilute aqueous solution or dispersion. Likewise, the fatty nitrogen compound and clay are also preferably added in the form of dilute aqueous solutions or dispersions. Where the fatty nitrogen compound is an amine, it can be solubilized by at least partially neutralizing the same with a water-soluble organic or inorganic acid such as acetic acid and mineral acids including hydrochloric, sulfuric and the like.

Using dilute solutions or dispersions of dye, clay and fatty nitrogen compound results in a more even distribution on the fibers. The solutions or dispersions should contain less than 10% by weight of these. It is often preferred that the alkali metal or acid montmorillonite clay be first added to the fibers followed by addition of the fatty nitrogen compound and then the anionic dye. In the preferred procedure where the anionic dye, fatty nitrogen compound and alkali metal or acid montmorillonite clay are thoroughly mixed with an aqueous pulp or fiber dispersion, sheets can then be prepared using conventional techniques.

Example 1: Use of Diamine – One-liter samples of an aqueous dispersion of moderately refined bleached kraft pulp at 0.8% solids at local tap water pH (7.9) were prepared. 20 cc of a 0.2% aqueous solution of Acid Violet were added to each sample. One sample was retained as a control; to the remaining samples were added various amounts of either or both of the following: a 1% aqueous solution of a diamine of the formula

$$R'\text{—}\overset{\displaystyle H}{\overset{|}{N}}\text{—}CH_2CH_2CH_2\text{—}NH_2$$

where R' is the mixed C_{14}, C_{16} and C_{18} carbon atom alkyl radical derived from the corresponding fraction of fatty acids obtained from tallow (the solution was prepared by adding 1 cc diamine per 100 cc H_2O containing HCl at a concentration of 1 cc/500 cc H_2O); and 1% suspension of degritted oil-well-grade Wyoming bentonite (the bentonite was degritted by centrifuging an approximate 5% slurry of bentonite to remove nonbentonite material and the supernatant was diluted to 1% solids with water).

The samples were stirred and formed into handsheets on a Noble and Wood 12-inch by 12-inch handsheet machine at the pH of local tap water. The handsheets were then dried at 200° to 250°F. The control containing no bentonite or diamine was not dyed. The sheet formed from a dispersion to which had been added 4 cc of the clay dispersion was very slightly dyed.

The sheets obtained using 1 and 2 cc of the diamine solution were lightly dyed. The sheets obtained using 4 cc of the bentonite dispersion and either 1 or 2 cc of the diamine solution were dyed violet with the latter being somewhat more deeply colored. This unexpected improvement due to the use of the bentonite clay is also evident as the amount of diamine used is increased although the degree of improvement is reduced.

Example 2: Use of Monoamine – The fatty diamine was replaced by a fatty monoamine of the formula $R''-NH_2$ where R" is hydrogenated tallow (a mixture of saturated radicals consisting principally of straight chain 16 and 18 carbon atom alkyls from the corresponding fraction of acids obtained from tallow). The dye used was a 0.2% aqueous solution of the acid dye Calcocid Blue 2G. Results are as follows:

Handsheet	Bentonite Dispersion, cc	Monoamine Solution, cc	Results
1	0	0	Not dyed
2	2	0	Not dyed
3	0	1	Slightly dyed
4	2	1	Dyed

Example 3: Use of Quaternary Compound – Example 2 was repeated except that the fatty monoamine was replaced by a quaternary compound of the formula:

$$\left[R''''-\overset{\overset{\displaystyle CH_3}{|}}{\underset{\underset{\displaystyle R'''}{|}}{N^+}}-CH_3\right]Cl^-$$

where R''' and R'''' are mixed C_{14}, C_{16} and C_{18} alkyl radicals as in the diamine of Example 1. Results are as follows:

Handsheet	Bentonite Dispersion, cc	Difatty Quaternary Solution, cc	Results
1	0	0	Not dyed
2	4	0	Slightly dyed
3	0	4	Dyed
4	4	4	Intensely dyed

Soluble Stilbene Dyes for Use in Manufacture of Colored Paper

In making colored paper the dyestuff should be in solution form so it can be metered into the stream of paper pulp immediately before the formation of paper on the web. The dyestuff must be rapidly and completely soluble in aqueous media in order that no undissolved particles remain to give a spotty appearance to the finished paper. One dyestuff widely used is Direct Yellow 11, C.I. 40000, which is one of the family of stilbene dyes prepared by condensing 2-methyl-5-nitrobenzenesulfonic acid in the presence of sodium hydroxide. *R.W.J. Rebhahn and D.W. Bannister; U.S. Patent 3,547,774; December 15, 1970; assigned to Toms*

Chemical Corporation provide this dyestuff and associated members of this same family in a liquid form which is readily soluble in water and by means of a process which does not require the separation of the dyestuff after its formation.

In the process 2-methyl-5-nitrobenzene sulfonic acid containing no more than 3% sulfuric acid is subjected to self-condensation in aqueous lithium hydroxide solution at a temperature within the range of about 40° to 100°C. The process is carried out employing the reactants in approximately three times the concentration conventionally employed in the preparation of stilbene dyes such as Direct Yellow 11 (C.I. 40000).

The process is usually complete in from about 4 to 8 hours. The dyestuff is obtained in the form of a liquid consisting of a saturated solution containing solid dyestuff and as such can be employed directly after the condensation reaction has taken place. No purification, filtration or drying of the dyestuff is necessary.

Example 1: Preparation of Dyestuff — A one-liter reaction vessel is charged with 100 g of water and with 150 g of 2-methyl-5-nitrobenzene sulfonic acid containing 2% sulfuric acid. The resulting mixture is heated to 55°C and adjusted to a pH of 7.5 with 32 g of lithium hydroxide. As soon as the solution of the 2-methyl-5-nitrobenzene sulfonic acid is complete there is added a further quantity of 30 g of lithium hydroxide monohydrate over a period of 20 minutes. The reaction mixture is maintained at 55°C for a period of 8 hours at a pH value in the range of 11.7 to 11.9.

At the end of the 8-hour period the volume of the reaction mixture is adjusted with water to 600 ml and stirred for a period of 12 hours. At the expiration of the 12-hour period the pH of the reaction mixture is adjusted to 8.0 employing 27.6 g of 96% sulfuric acid. The reaction mixture is then adjusted to a volume of 625 ml and employed directly in the coloring of paper according to the procedure set forth in Example 2.

Example 2: Use in Manufacture of Colored Paper — Five parts of bleached sulfite pulp are immersed in 200 ml of water at room temperature and the pulp suspension agitated until the pulp is thoroughly dispersed. 1 ml of 5% rosin solution is then added and the resulting mixture agitated for 10 minutes.

An amount of the dye solution of Example 1 sufficient to provide a concentration of 2.0% dye on the weight of the dry bleached sulfite pulp is added and the suspension again agitated for 10 minutes. 3 ml of 5% aluminum sulfate solution are then added, the suspension agitated for 10 minutes and sheets then prepared in a hard-sheet mold. The sheets with blotters are pressed in a Williams standard pulp press at a pressure of 1,000 pounds for a period of 80 seconds. The sheets are then removed with blotters and dried.

The above procedure is repeated employing an amount of dyestuff solution from Example 1 sufficient to provide a concentration of 0.4% of dyestuff on the weight of the dry bleached sulfite pulp. In each instance colored paper of satisfactory yellow shade is obtained.

Liquid or Granulated Dye Preparation

A. Frei; U.S. Patent 3,894,836; July 15, 1975 is concerned with a process for dyeing paper, which is characterized by the use of the dye of the formula

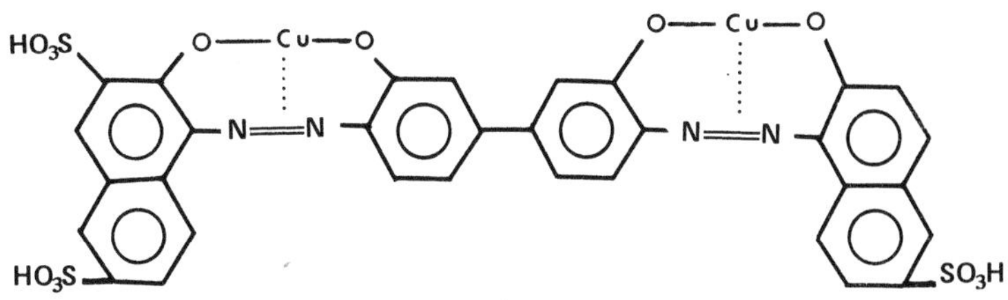

The dye is employed preferably as one of its salts, for instance an alkali metal (sodium or potassium) salt or an ammonium salt. The dye is known and is used for dyeing cotton, on which, as is stated in the literature, it requires heating for fixation.

It has been found that the dye of the formula above is eminently suitable for paper dyeing. It can be used for dyeing unsized and sized (previously or subsequently sized) paper grades by dipping, coating, spraying or other suitable methods, though it is especially valuable for application in the stock prior to sheet formation.

The dye, which is well soluble in water, gives level dyeings of a brilliant reddish blue shade on paper which are notable for their fastness; they are very fast to light, fading tone-in-tone after a long period of exposure, and show similarly good wet fastness properties (alcohol, milk and in particular water fastness).

The dye of the above formula can be converted into liquid dye preparations containing 1 part by weight of dye, 3 to 11 parts by weight of water and 0.5 to 3 parts by weight of an amino alcohol, preferably a lower mono-, di- or trialkanolamine in which each alkanol radical optimally contains 2 or 3 carbon atoms.

These liquid preparations are produced by mixing the stated components, if necessary with heating for a short time; they may contain further additives, such as wetting or dispersing agents.

The new liquid preparations are true solutions or, if concentrated, dispersions (suspensions). They can be employed for dyeing as such or after dilution with water as stock solutions. They are excellent for preparing liquors and pastes for paper dyeing by coating, spraying and dipping methods; however, they are especially well adapted for addition to the stock before conversion into sheet, where they can be metered continuously into the stock stream in the undiluted state.

The dye preparations have the advantage of being easier to handle. The inconvenient dusting that occurs when powder dyes are weighed out is absent. They are storable at room temperature for several months, withstand frost and are immune from mold growth.

The dye of the formula above can be employed with equal success in the form of granules, preferably of neutral to alkaline reaction, instead of the liquid form.

The granulated preparations of the dye are characterized by containing, with the dye, a maximum of 50% of a salt (commonly used in dyeing preparations) which dissociates in water and/or a nonhygroscopic dextrin; the average size of the granules is at least 20 microns. Salts of neutral or alkaline reaction which dissociate in water are suitable, but preference is given to the salts of alkali metals, e.g., sodium and potassium chloride, sodium and potassium sulfate, monosodium and disodium phosphate, sodium and potassium pyrophosphate, sodium and potassium carbonate. Nonhygroscopic dextrins which form no lumps when stirred in aqueous medium are especially suitable.

The granulated preparations can be produced with the dye in the form of the moist acid presscake of low salt content (preferably not more than 25% of its dry weight of inorganic salts) as obtained in manufacture after acidification. As the granules are best produced by granulation, for example atomization of a neutral or alkaline liquid preparation, if an acid presscake is used, it is best to neutralize it with a strong base, such as lithium, sodium or potassium hydroxide.

The granulated dye preparations can be produced by dissolving or suspending the dye, the salt or salts, the dextrin, and (if used) the strong base in two to ten or preferably two and a half to five times the amount of water. If a suspension is formed, it is advantageously homogenized. The solution or suspension is granulated by one of the standard methods, preferably atomization, to give granules averaging at least 20, preferably 40 to 120 microns in size.

The granules are free-flowing and readily soluble in water so they are excellent for dry addition to paper stocks. They also show good storage stability, especially chemical stability. Compared with the powder dyes used to date they offer the advantages of better wettability and higher solubility in water, this being especially true of preparations containing a maximum of 30% of a salt which dissociates in water.

This means that stock solutions of concentrated strength can be prepared. Moreover, the granules are easier to handle in weighing out additions for dyeing, and for loading into containers.

Example 1: Liquid Form — One part of the dye of the formula on the preceding page in the form of a moist mineral acid presscake of 25% strength is dissolved in 3.2 parts of 70% diethanolamine solution. This preparation shows good stability on storage for several months at room temperature.

One hundred parts of chemically bleached sulfite pulp suspended in water are beaten to the desired fineness; 3 parts of the preparation described above and, after 15 minutes, the sizing agent (which fixes the dyeing) are added. Paper made from this stock is dyed to a reddish blue shade of medium depth which has good fastness properties. The white water from the paper machine is colorless.

Example 2: Granulated Form — Fifty-three parts of the dye of the formula on the preceding page in the form of the moist acid presscake obtained on manufacture, which contains 20% dye, 5.5% inorganic salts and 74.5% water, are added to 300 parts of water, along with 5.5 parts of lithium hydroxide and 20 parts of dextrin. Stirring gives a homogeneous suspension which is converted in an atomizer drier into granules averaging 40 microns in size. These are readily soluble in water and very suitable for paper dyeing.

One thousand parts of chemically bleached sulfite pulp in aqueous suspension are beaten to the required fineness, 0.5 part of the granulated dye preparation above is added, followed after 15 minutes by the sizing agent, which fixes the dye. The paper made with this stock has a reddish blue shade of medium depth showing good fastness properties. The white water is colorless.

Process for Obtaining Multicolored Effect on Paper

M.J. Plumez, C.B. Brown and W.S. Karro; U.S. Patent 3,562,098; February 9, 1971; assigned to Geigy Chemical Corporation describe a process for obtaining such special effects on paper as multicolored spotted, splashed, bleeding madras, and hombre effects.

This comprises adding to a cellulose substrate, in particular paper, a colorant such as a dye or pigment, which colorant is insoluble in water, but soluble in organic solvents, and during or after the drying of the paper to which the colorant has been added, treating the paper with an organic solvent in which the colorant is soluble.

The colorant, which term is also intended to include mixtures of different colorants, may be applied to finished sheets of paper from an aqueous dispersion system in any conventional manner as for example in a size press, coater or by means of a pickup roll. The sheet is then dried, and either after or during the drying operation the sheet is treated by means of spraying or coating, with an organic solvent to dissolve at least a portion of the colorant and thus produce a multicolored effect.

The effect achieved of course will vary according to whether a coating of solvent is employed, or if the solvent is sprayed, splashed, or otherwise differentially applied to areas of the paper surface. The particular effect achieved in any one instance of course will depend to a degree upon the kind and amount of colorant present, and the relative speed and degree of solubility of the colorant in the organic solvent.

The colorant employed alternatively may be added to the paper by addition to the pulp prior to the formation of the sheet. In this case the colorant is added to the pulp mixture in dry powder form. It is preferred that the pulp mixture contain a sizing agent such as rosin or a dispersion agent such as for example a condensate of naphthalene sulfonic acid and formaldehyde as disclosed in U.S. Patent 1,336,759.

Generally about 1% by weight of rosin or dispersant has been found to be satisfactory although the amount is not critical and the use of the rosin or dispersant, while preferred, is not absolutely necessary. The pulp is formed into the paper sheet in a conventional manner, and during or after completion of drying the sheet, the organic solvent is applied to at least partially dissolve and develop the colorant.

Among the solvents which are most often employed are methyl and ethyl alcohol, cyclohexanone, benzyl alcohol, and methylcellulose. The dyes are generally employed in amounts of from about 0.01 to 10% by weight of the dry pulp.

Example 1: Colorant Added to Pulp Slurry – 80 g of dry sulfite pulp is slurried in a solution of 7.2 g/l of alum (17% assay of Al_2O_3) to a total volume of 4 liters. To the slurry is added 0.8 g of Solvent Violet 11 (C.I. 61100) and 1.2 g of rosin and the slurry agitated for 5 to 10 minutes.

The pulp slurry is diluted to a consistency of 0.4% and a handsheet formed by suction on an 80 mesh copper screen. The sheets are pressed and dried in a rotating drum dryer. The sheets are then sprayed with ethyl alcohol and sheets dried of the solvent alcohol. The resulting sheets exhibit a very attractive spotted violet appearance.

The above procedure is repeated but employing Solvent Red 36 (C.I. 13900) in place of the Solvent Violet 11 and using benzyl alcohol as the solvent. A pleasing spotted red colored paper is obtained.

Example 2: Colorant Added to Paper Sheet – Paper sheet is formed as in Example 1, except that the dye is not added to the pulp slurry. After the paper sheet is formed and dried, an aqueous dispersion of 1% by weight of Solvent Violet 11 and the naphthalene sulfonic acid-formaldehyde condensation product of U.S. Patent 1,336,759 is applied to the surface of the paper sheet by means of a pickup roll and the sheet again dried. The sheet is then sprayed with solvent as in Example 1 with generally similar results being obtained.

Surface Coloring Using Slit Aperture Method

B.A. Wennerblom and G.R. Svensson; U.S. Patent 3,941,902; March 2, 1976; assigned to Svenska Cellulosa A/B, Sweden describe a method of continuously making surface-treated paper in which the web is treated with agents (dissolved or dispersed in water) which are found to be suitable for surface treatment purposes.

As shown in Figure 2.1a, the surface treatment proper is to be carried out with a coating apparatus, which briefly can be called a slit unit, because it is characteristic of this unit that it comprises a slit-shaped aperture **2** in and along an oblong container **1**, which is placed across the running direction of the paper web **3**, and to which the treatment liquid is supplied. The container **1** in its most simple design is a straight pipe. The sizing liquid can be supplied either via one end of the pipe or via one or more connections along the pipe (not shown).

The slit-shaped aperture **2** in the container **1** extends along the longitudinal direction of the container. The length of the aperture **2** corresponds to the width of the paper web to be treated. As no liquid must leak out outside the paper web edges, the slit aperture **2** must be at least about 1 cm shorter than the web width.

The uncoated edge strips of the web may, where occurring, be cut off in connection with the reeling operation. The slit aperture should have a width of 0.05 to 5 mm, preferably 0.1 to 2 mm.

The angles α and β formed by the paper web with the outer surfaces defining the slit aperture **2**, as indicated in Figure 2.1b, are of importance for the process and should suitably be between 0° and 45°, preferably between 5° and 30°. In special experiments it was surprisingly found essential for the smoothness of the application, that the paper web form a clearance angle to the rear surface of the slit, seen in the running direction, which is not too small. The angle α thus should be at least 30°, suitably at least 90° and preferably at least 120°.

FIGURE 2.1: SURFACE COLORING USING SLIT APERTURE METHOD

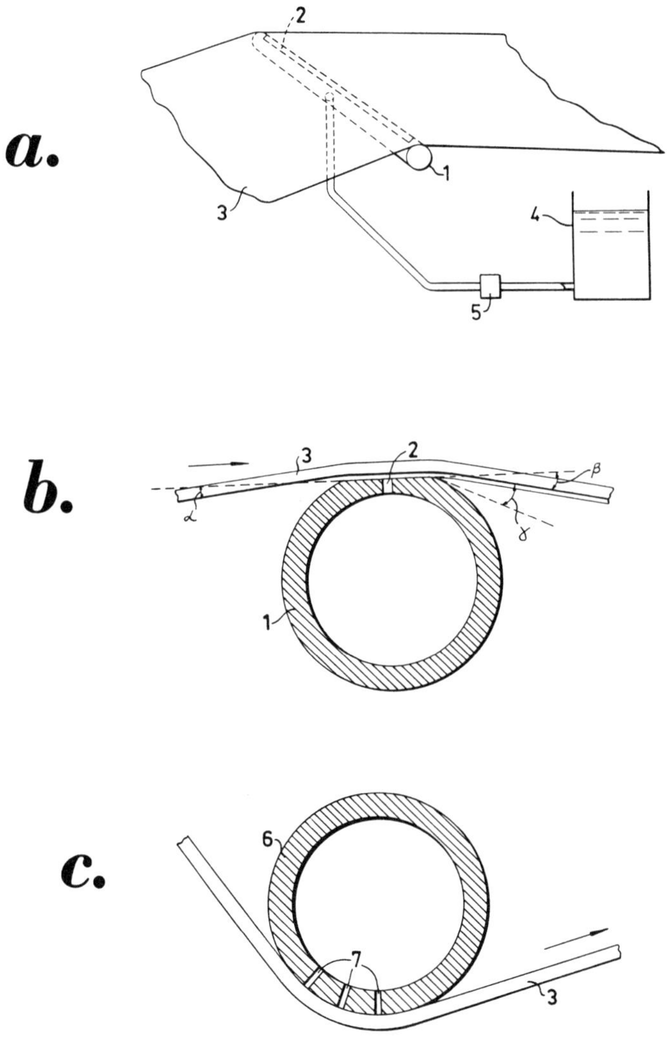

Source: U.S. Patent 3,941,902

The paper web, however, owing to imperfections in the function of the paper-making machine, often has nonuniform properties both across and along the web. The tension in the web, therefore, is not entirely uniform across the web and can in a given position also vary with the time.

In order to ensure in such cases that the slit abuts the web uniformly, it can be necessary to arrange the slit pipe so as to be easily flexible and adapted to be pressed against the paper web with a constant or substantially constant pressure. The elastic pipe is then to be supported on a structure resistant to bending and be pressed against the paper web by means of, for example, adjustable screws, springs or hydraulic pressure.

The same effect can be achieved with a slit container resistant to bending when the tension of the paper web is controlled by sliding or rolling means, the pressure of which against the web can be adjusted over the width. This pressure can also be affected by air, for example from a stationary pipe **6** shown in Figure 2.1c.

The pipe **6** extends across the width of the paper web **3** and is along its length and over the circumferential portion being enclosed by the paper web **3** provided with apertures **7** in the form of holes or slots, through which the air from inside the pipe **6** is forced out between the pipe and the paper web **3**.

Web portions with a tension lower than in the main part of the web are thereby pressed by the air farther outwards from the pipe, thereby balancing the difference in pressure. This compressed air device can be placed immediately before or immediately after the nip, seen in the running direction of the paper web.

Example 1: Two experiments with two-side surface coloring were carried out. The base papers used were two periodical print papers with the following characteristics:

	A	B
Substance, g/m^2	51	67
Fiber composition:		
Mechanical wood pulp, %	85	87
Chemical pulp, %	15	13
Ash content, %	7	12
Moisture content, %	6.8	7.2
Surface roughness, Bendtsen ml/min	230	500

Basic coloring agents were used as the coloring fluid. The coloring solutions were prepared by mixing the fluid coloring agent with water. For preparing a special yellow and a special green color tone, the following coloring agent combinations were used:

	A	B
Auramin fluid conc., %	0.05	0.04
Chrysoidine, %	0.10	-
Burma green, %	-	0.04

The surface coloring was carried out in a half-scale machine, in which the paper at first was treated on the wire side and then on the upper side. The solution at room temperature was supplied continuously to the slit. The machine speed was set so that the amount applied exceeded with a good margin P_{min} (the smallest uniformly applicable amount of solution) in order to obtain a deposit being equally large on the wire and on the upper side, and thereby also to obtain the same color intensity on both sides.

The amount applied in this experiment was 10.0 cm^3 on the respective sides. The treated paper was dried with hot air. In both experiments the papers became colored very uniformly.

Example 2: For experiments with one side surface coloring without subsequent drying, a super-calendered paper with the substance 55 g/m^2 was used, which was made of 25% chemical pulp and 75% mechanical woodpulp. The ash content of the paper was 10% and the moisture content 6.3%. The surface roughness of the upper side was 50 ml/min according to Bendtsen.

As coloring agent fluid basic coloring agents from BASF were used. The coloring solution was prepared by mixing 2.6% Auramin fluid concentrate and 0.1% Burma green with water.

The surface coloring was carried out in a high-speed machine. The paper was treated on the upper side. The solution at room temperature was supplied continuously to the slit, and the flow was so adjusted that the amount applied was immediately above P_{min}. The machine speed was 170 m/min. The amount of coloring solution applied was 3.7 cm^3/m^2.

The colored paper was reeled without drying. The moisture content in the base paper was 6.3%, and after coloring and reeling 10.1%. The calculated moisture content after the coloring is 13.0%. The moisture content in the paper, thus, is reduced by 2.9% in the passage between slit and reeling, due to natural evaporation. The paper was colored very uniformly and could be used in its prevailing state, without further drying. It was adapted, for example, to be used for placards.

ORNAMENTAL PAPERS

Ornamental Paper Incorporating Iridescent Plastic Film Particles

H.A. Walters and R.M. Gooch; U.S. Patent 3,442,755; May 6, 1969; assigned to The Dow Chemical Company provide an attractive, eye-catching paper having intimately incorporated therein a plurality of thermoplastic resinous elements, the resinous elements each comprising a transparent thermoplastic resinous body composed of a plurality of generally parallel layers of transparent thermoplastic resinous material.

At least about 20% of the layers have a thickness of between about 0.05 to 5 microns and preferably 0.05 to 1 micron for maximum iridescent effects, and differ in refractive index from adjacent layers by at least about 0.03.

The iridescent synthetic thermoplastic resinous elements are readily prepared in the form of thin sheets or films beneficially having thicknesses from about 0.25 to 10 mils by the simultaneous extrusion of a plurality of layers of diverse thermoplastic resinous materials and subsequently stretching the extrudate at a thermoplastic temperature until the thickness of the composite sheet and the layers therein is reduced to the desired level.

Beneficially, such a composite sheet can be obtained by the simultaneous extrusion of a plurality of layers from a single die or by preparing a composite sheet by hot lamination of film or extrudate to provide the desired number of layers, heating the composite sheet to a thermoplastic temperature and reducing its thickness by stretching.

Once the composite iridescent film is prepared, it is then cut or processed into the desirable shapes such as elongated strips, rectangles, circles and even chopped into random shapes, depending upon the desired effect. Oftentimes particularly attractive effects are obtained by using film having layers of various thicknesses so that the iridescent colors obtained are varied.

In order to achieve a satisfactory iridescent effect, the layers of material must be deposited in such a manner that a relatively sharp interface is maintained. Such an interface can be maintained by the use of synthetic resinous materials in solvents which are solvents only for one of the materials.

Thus, distinct layers are obtained having a sharp interface and as successive layers are deposited the iridescent effect increases. About 10 layers are satisfactory for minimal iridescent effect. However, 50 is usually the number of minimum layers suitable for a relatively intense iridescent effect and desirably, composite structures having 100 to 500 layers are employed for maximum iridescent effect.

Example 1: A paper is prepared employing an iridescent film which is 1 mil thick comprising 125 layers of polystyrene and polymethyl methacrylate alternately disposed with respect to each other wherein the thickness of the polymethyl methacrylate layers was about 3 microns. The film was placed in water and shredded by means of a high-speed mixer to provide a slurry of about 0.5% by weight solids. A bleached kraft pulp was beaten to a Canadian Standard Freeness of 397 ml. Ten parts by weight of pulp based on the dry weight and 1 part of the shredded foil were mixed and added to the deckel box of an apparatus adapted to prepare a handsheet.

The water drained through an 80 mesh screen from the deckel box and the pulp and foil were retained on the screen. The resulting handsheet and screen are removed from the deckel box, the sheet pressed against and transferred to a sheet of blotting paper and the handsheet then placed against a polished chromium-plated sheet under a pressure of about 60 pounds per square inch for a sufficient length of time to remove by capillary action the majority of the water available to the blotting paper. The sheet of blotting paper is removed and replaced with ½-inch-thick felt.

The felt-paper-plated sheet sandwich is placed in a platen press between closed platens for a period of 4 minutes and heated to reduce the water content of the sheet to about an equilibrium value at room temperature. The resultant sheet showed a plurality of fibers dispersed on and about the surface which had an attractive iridescent appearance and on movement of the sheet, the fibers appeared to change color and appear and disappear. Flexing of the sheet would not dislodge the iridescent film particles and the attractive appearance was maintained.

Example 2: (a) In a manner similar to the procedure of Example 1, a handsheet was prepared employing equal parts of pulp and iridescent film with the exception that a portion of the pulp was deposited on the screen before the addition of the iridescent film particles to the deckel box.

This resulted in a sheet having iridescent film deposited on one side rather than on both sides as in the previously described procedure. The paper showed a considerably greater iridescent effect.

(b) A handsheet was prepared employing a black pulp employing 10 weight percent iridescent film and a startlingly attractive iridescent sheet was obtained. Repetition of the foregoing procedures utilizing iridescent film in the form of filaments about 1/16 of an inch in width and about 1 inch in length, film in squares of about 5/12 of an inch per side and discs of about 1/8 inch in diameter also result in attractive and eye-catching iridescent paper.

Similar beneficial results are obtained when a pulp is replaced with cotton fibers, fur, asbestos, glass and the like. Excellent iridescence is also achieved when a pulp is replaced with cotton fibers, fur, asbestos, glass and the like. Excellent iridescence is also achieved when laminates are prepared of such materials as polyvinyl chloride-polystyrene, polystyrene-polyethylene, polypropylene-polystyrene, ethylcellulose-styrene, and the like.

Laminated Paper with Dispersed Decorative Elements

F.S.T. Fung and R.H. Doughty; U.S. Patent 3,451,878; June 24, 1969 provide a decorative laminated paper having decorative elements dispersed therein at various depths to produce a three-dimensional effect and to vary the color effect of like-colored particles.

The process is described with reference to Figure 2.2a through e. Figure 2.2a illustrates an enlarged plan view of a decorative element; Figure 2.2b shows an enlarged sectional view of a decorative element **1** taken through **2–2** of Figure 2.2a; Figure 2.2c shows an enlarged cross-sectional view of paper sheet having individual decorative elements distributed substantially uniformly throughout; Figure 2.2d illustrates an exploded view of the laminate containing a print sheet of the type illustrated in Figure 2.2c; and Figure 2.2e shows an exploded view of a laminate containing an overlay sheet having individual decorative particles distributed substantially uniformly throughout.

The decorative particle shown in Figure 2.2a is a circular shape. However, it may be of any desired shape such as hexagonal, rectangular, square, etc. As shown in Figure 2.2b, decorative element **1** comprises a paper base **2** having an ink coating **3** thereon. The coating **3** penetrates the paper base **2** only to the extent required for proper binding leaving the remainder of paper base **2** unsaturated.

The coating **3** may be any well-known commercially available high solids laminating grade inks which resist change in color or stability in laminating processes. The extent of penetration by coating **3** will depend upon the coating weight and structure of paper base **2** and more particularly, the porosity of the paper base **2**. It has been found that papers having a Gurley densometer value in the range of less than 70 seconds and greater than 20 seconds for 400 ml of air are of suitable porosity. The coating weight should be no greater than 15 lb/3,000 ft². By keeping the coating weight and the porosity of base paper within the foregoing ranges, the coating **3** penetrates the base sheet to a sufficient extent for binding purposes but not so completely as to completely saturate the base sheet. This leaves sufficient area in the finished coated paper available for penetration by a thermosetting resin.

FIGURE 2.2: LAMINATED PAPER WITH DISPERSED DECORATIVE ELEMENTS

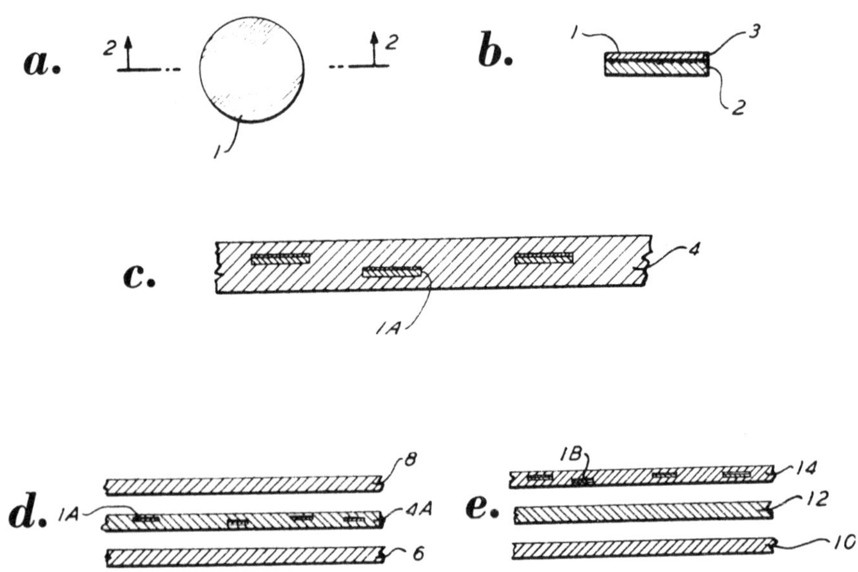

Source: U.S. Patent 3,451,878

The inclusion of foreign particles, which are not capable of absorbing the thermosetting resin, in lamination layers is responsible for blisters in laminates. The laminate fails by delamination in the area of the particles. Since the decorative dots of the process are capable of absorbing the resin, the blistering possibility is greatly decreased.

Referring to Figure 2.2c, decorative elements of the type shown in Figure 2.2b are shown at **1A** dispersed within a colored or otherwise decorated paper sheet **4** commonly referred to as a print sheet when employed in a paper laminate as shown in Figure 2.2d. This laminate comprises a core sheet **6**, a print sheet **4A** and an overlay sheet **8**.

Core sheet **6** comprises an unbleached kraft paper impregnated with a thermosetting resin such as phenol-formaldehyde resin. The overlay sheet **8** and print sheet **4A** are made of paper comprising alpha-pulp, rayon or similar natural or synthetic fibers or mixtures thereof. The overlay sheet becomes transparent or translucent when impregnated with resin and cured. Print sheet **4A** is manufactured with the inclusion of an inorganic pigment or mixtures of inorganic pigments making possible a wide variety of colors.

The pigments employed are well known and include, e.g., titanium dioxide (white), cadmium sulfide (yellow), etc. The greater the amount of pigment in the print sheet, the greater the amount of light is blocked out as the distance of the decorative element **1A** is increased from the surface of the sheet **4A**. Due to the partial hiding power of print sheet **4A** and due to the various depths at which the particles are disposed, the inclusion of the decorative elements **1A** in sheet **4A** produces a three-dimensional effect.

As shown in Figure 2.2e, the decorative particles may be included in the overlay sheet of a paper laminate. This laminate comprises a base core sheet **10**, a print sheet **12**, and a transparent overlay sheet **14** having decorative particles **1B** dispersed therethrough.

The overlay sheet **14** is oversaturated with a thermosetting resin to get good protection against wear. By oversaturating the overlay sheet, all available areas in the decorative particles **1B** will be saturated with thermosetting resin, thereby greatly minimizing the possibility of delamination to an even greater extent than when the decorative particles are included in the print sheet.

The lower print sheet **12** in the laminate cannot be oversaturated as that would result in a laminate having poor mechanical properties due to the presence of excessive amounts of the hard and brittle thermosetting resin in the body of the laminate.

SECURITY PAPERS

AUTHENTICATABLE PAPERS

Paper Containing Azoic Coupling Component

It is frequently desirable to have paper, such as used for legal documents, bank notes, checks and other security documents, which can be readily identified or authenticated by the user to enable him to ascertain its genuineness. To detect forgeries, the genuine paper should have a feature or characteristic easily detectable, for example by a bank cashier, but which cannot be reproduced without great difficulty.

W.W. Krueger and J.F. Dever; U.S. Patent 3,523,866; August 11, 1970 describe an authenticatable paper containing an azoic coupling component which, upon treatment with a solution of an azoic diazo component, produces an insoluble azo dye as an identification color suitable for authenticating the paper.

The azoic coupling components suitable for use belong for the most part to the Naphthol AS class of coupling components. The structures of such compositions are set forth in the Colour Index, Society of Dyers and Colorists, Bradford, England, and the American Society of Textile Chemists and Colorists (second edition, 1956, and supplements thereto), and they are identified by the term C.I. azoic coupling component, which is usually followed by a number to identify specific compounds within the class.

The preferred C.I. azoic coupling components have C.I. numbers of C.I. 37505 and higher, up to C.I. 37625. The azoic coupling components within this series of C.I. numbers include the anilide of 3-hydroxy 2-naphthoic acid (Naphthol AS, C.I. 37505) followed by other arylamides in increasingly complex chemical progression, those having higher C.I. numbers comprising more complicated derivatives based on the acetoacetic esters with other derivatives such as benzoyl-acetic acid esters, benzocarbazoles and dibenzofurans.

Unless specially prepared by the manufacturer, a C.I. azoic coupling component is insoluble in water and is suitably converted to the water soluble salt by dissolving the coupling component in an alkaline solution, e.g., caustic soda, prior to being applied to paper.

The paper to be treated with the alkaline solution of a C.I. azoic coupling component can be made on any suitable papermaking machine, including Fourdrinier and cylinder machines. The alkaline solution of the azoic coupling component can be added to the paper at any one of a number of stages in the papermaking process. The alkaline solution can be added to an aqueous slurry containing the pulp before it is fed to the papermaking machine; when added to the pulp slurry, the azoic coupling component becomes impregnated throughout paper made from the slurry.

The azoic coupling component can be added at the dry end of the papermaking machine by application at the size press, in which case the coupling component is contained both on the surface of the finished paper and partly impregnated within it. Further, the azoic coupling component can be applied to the paper by printing, in which event it is contained on the surface of the finished paper.

When it is desired to produce an authenticating or identifying azoic color with the paper containing a C.I. azoic coupling component, the paper is treated with a solution of a diazotized primary amine to form an insoluble azo dye. In the Colour Index, the primary amines that comprise azoic diazo components are identified and referred to by a C.I. number, e.g., C.I. 37005, as well as by the term C.I. azoic diazo component which is usually followed by a number such as, for example, C.I. azoic diazo component 2. The diazo components commence with C.I. 37000 and proceed to C.I. 37275. Such numbering begins with the derivatives of aniline, followed by derivatives of toluidine, anisidine, etc. and building up to the more complicated structures for compounds having higher numbers.

Azoic diazo components are purchased as stabilized diazo compounds generally referred to as salts or fast color salts. These salts need only dissolving in cold water to be ready for use, thus eliminating the process of diazotization, and are preferred because of their convenience of use by the person authenticating the paper.

For authentication of paper containing a C.I. azoic coupling component, an aqueous solution of a fast color salt or other form of the azoic diazo component is prepared and applied to the treated paper by any suitable applicator. The fast color salt solutions should be used within 24 hours after preparing since they will become unstable and ineffective with the passage of time. An insoluble azo dye will then form in situ on such contacted portion of the paper providing a predetermined, distinctive azoic coloring that enables the authenticity of the paper to be established.

An azoic coupling component concentration in the range of about 0.25 pound (of unreacted coupling component) per ton of paper, and above, is capable of producing an authenticatable color change upon contact with a diazo component. The concentration of the diazo component solution used for authentication will also affect the shade or intensity of the color that is produced; solutions on the order of 0.5% by weight and above are satisfactory for obtaining an authentication color, with a 1.2% by weight solution generally being capable of reacting with all of the coupling component added to the paper at the levels used in the examples. Because the coupling component can be put in or on the paper without significantly changing either the base color of the paper or its visual appearance, the latently authenticatable paper can have the customary appearance of the usual white paper. However, the paper can also be partially or completely colored with an azoic color, but have unreacted coupling component available for authentication. Also, once the coupling component has been applied to the paper, the paper can then be printed upon and the azoic color produced will still appear around the printed matter.

Another advantage of the coupling components used herein is that with a single coupling component a multiplicity of predetermined azoic color combinations can be provided merely by contacting such component with a different particular C.I. azoic diazo component. The range of authenticating colors can be further extended by using mixtures of the coupling components in the paper, and/or mixtures of the azoic diazo components for the authenticating solutions. The wide variety of specific color reactions provided with the authentication papers has the advantage of enabling a particular color reaction to be reserved for a specific user so that only he will know the authentication color that is to be produced; this also enables different authenticating colors to be used at different times should a time-coded authentication feature be desirable.

Example: (a) — A slurry was made up of 100 parts by weight of pulp, comprising 60% hardwood sulfite, 30% refined kraft and 10% broke, to which the following were added: alum at a concentration level of 40 pounds per ton of pulp; rosin at a concentration level of 15 pounds per ton of pulp; hydrated aluminum oxide at a concentration level of 150 pounds per ton of pulp; and titanium dioxide at a concentration level of 70 pounds per ton of pulp.

(b) — To a portion of the above stock slurry, there was added a caustic solution of C.I. azoic coupling component 4 (C.I. 37560; Naphthol AS-BO) at a concentration level of 10 pounds of the coupling component per ton of slurry, the ambient pH being 4.7. The slurry was then fed to a papermaking machine, and the finished dried paper was subjected to the following authentication tests.

Authentication Tests — Aqueous solutions of a number of azoic diazo components in the form of fast color salts were prepared and applied to selected paper specimens. In each instance an authenticating color was produced comprising an insoluble azo dye, the resulting azoic color in each case being indicated in the following table.

Fast Color Salt	C.I. Azoic Diazo Component Number	C.I. Number	Authenticating Azoic Color
Fast orange salt GGD	16	37045	Tan
Fast scarlet salt R	13	37130	Rust
Fast red salt GG	37	37035	Tan
Fast red salt B	5	37125	Russet
Fast bordeaux salt GP	1	37135	Light burgundy
Fast garnet salt GBC	4	37210	Burgundy
Fast blue salt BB	20	37175	Light blue

(c) — To a portion of the above stock slurry, there was added a caustic solution of the same coupling component as in (b), the pH, however, being 7. The slurry

was then fed to a papermaking machine, and the finished dried paper was subjected to authentication tests in a manner similar to (b), the resulting azoic color in each case being indicated in the following table.

Fast Color Salt	C.I. Azoic Diazo Component Number	C.I. Number	Authenticating Azoic Color
Fast orange salt RD	49	37050	Pink
Fast scarlet salt G	12	37105	Rose
Fast red salt GL	8	37110	Burgundy
Fast red salt AL	36	37275	Tan
Fast bordeaux salt BD	40	37170	Light burgundy

Paper Containing Ultraviolet Inhibitor

B.D. Skofronick; U.S. Patent 3,464,841; September 2, 1969; assigned to Custo-mark Corporation provides that a security impression mark which is normally invisible on a surface sheet of paper or in the matrix of the paper is rendered visible by wetting the paper with water. In the wet state, the previously invisible security mark is rendered visible, and again becomes invisible in the dry state. These security marks can be repeatedly made visible over extended periods of time by rewetting the paper. The security impression mark is ink receptive, by which is meant that it receives printed or other ink impressions without blurring, feathering or skipping. In other words, the security impression can receive normally visible written, drawn or printed impressions, including the nonink type, such as pencil or crayola.

The security impression mark or configuration or design is laid down or applied to a portion of the surface of finished paper. The security impression mark is laid down from a liquid composition which contains one or more ultraviolet absorbent organic chemicals and an organic solvent which is removed after application by evaporation, heat or the like. In a sense, the security impression may be said to be printed where a stamp or die is used to impart the design or configuration onto the hard finished surface. The liquid composition may be spread as a film on such impression transferring means and applied with momentary pressure. The security impressions could be continuously impressed on papers in large scale operations by employing a printing roll or printing plate or a stencil operation or similar means for large scale operations.

The ultraviolet absorbing organic chemicals which are useful are known and represent various benzophenone types, particularly hydroxy and alkoxy types of substituted benzophenones, among which are 2-hydroxy-4-methoxybenzophenone (Uvinul M-40), ethyl-2-cyano-3,2 diphenyl acrylate (Uvinul N-35), 2-hydroxy-4-methoxybenzophenone-5-sulfonic acid (Uvinul MS-40) and 2,4-dihydroxybenzophenone (Uvinul 400). The Uvinul ultraviolet absorbent organic chemicals are made by Antara Chemical Company.

Other available and useful ultraviolet absorbing organic chemicals are the series supplied under the designation of Eastman Inhibitors by the Eastman Chemical Company. They are HPT—hexamethylphosphoric triamide, DHPB—2,4-dihydroxybenzophenone, OPS—p-Octylphenyl salicylate, RMB—resorcinal monobenzoate and DCBP—4-dodecyloxy-2-hydroxybenzophenone. Other ultraviolet absorbing chemicals commercially available on the market are Cyasorb UV9—2-hydroxy-4-methoxybenzophenone and Agent AT-539 (Uvinul N 539)—2-ethylhexyl-2-cyano-3,3 diphenyl acrylate.

The Eastman inhibitor HPT is supplied as a liquid containing organic solvent and may be applied in such a form directly to the surface of the finished paper and dried. The OPS and RMB have ester linkages but are believed to convert to hydroxybenzophenones under the influence of ultraviolet rays.

In general, a 10% concentration of the organic chemical and organic solvent will provide an operable liquid composition which leads to satisfactory security impression marks. This concentration can be raised to about 20 or 30%, depending on the degree of solubility of the chemical in a particular solvent, and the viscosity of the resulting liquid composition, which may affect handling.

It has been found that a large variety of these ultraviolet absorbent organic chemicals may be used to make the security impressions, so long as the chemical is compatible with the paper for the intended purpose, that is, it dries to an invisible or virtually invisible mark and does not discolor or provide any substantial telltale signs in the dry state. Likewise, the chemical must be compatible in the sense that it is ink-receptive and preferably useful over extended periods of time so that it can be subjected to repeated steps of exposure upon wetting, disappearance upon drying and reexposure upon rewetting.

A large variety of organic solvents can be used to prepare the liquid composition with the ultraviolet absorbing chemical. The solvent selected has only a few requirements to meet operability, namely, its ability to dissolve the ultraviolet absorbing organic chemical, its volatility so as to allow convenient removal after application by evaporation, heat, passage of warm air or the like, and its compatibility with the paper. By compatibility is meant that the organic solvent does not adversely affect the paper by discoloring it or leaving other telltale signs after removal following application. Likewise, the organic solvent should not damage the paper by any substantial weakening of the fibers, caustic action or the like.

Example: A representative number of experiments are summarized wherein various ultraviolet absorbing chemicals in a liquid composition are applied to the surfaces of finished papers. The organic solvent liquid composition is methyl Carbitol (diethyleneglycol methyl ether) and the ultraviolet absorbing organic chemicals are present therein, in a concentration of about 10% by weight. Such a liquid composition is contacted with an elastomeric hand stamp, and the wetted hand stamp is then contacted with a portion of the surface of the finished paper. The security impression is allowed to dry by evaporation of the organic solvent.

All of the security impressions in the dry state were invisible, or virtually invisible, and were distinctly visible after wetting the paper with water. The results of artificially aging the papers with the security impression marks are summarized in the following table. The indications under the Fadeometer heading report the appearance of the rewetted mark after 6 hours in the Fadeometer. The Fadeometer is a well-known instrument for comparing the fading properties of different materials which are exposed under standard conditions of humidity and the like to a carbon arc which operates as artificial sunlight.

The designations under the column entitled Oven Age report the appearance of the mark after the paper has been rewetted following 72 hours in an oven at 105°C. In all instances, the mark has excellent ink receptivity in that none of the ink feathers or blurs on the portion of the paper where the security impression has been applied.

Material Tested	Security Mark's Visibility Fadeometer	Oven Age
Eastman inhibitor HPT	Visible	Visible
Eastman inhibitor DHBP	Visible	Visible
Eastman inhibitor OPS	Slightly visible	Slightly visible
Eastman inhibitor RMB	Visible	Visible
Uvinul M-40	Slightly visible	Slightly visible
Uvinul N-35	Slightly visible	Very slightly visible
Uvinul MS-40	Slightly visible	Slightly visible
Uvinul 400	Visible	Visible
Agent AT-539	Slightly visible	Visible
Cyasorb UV-9	Slightly visible	Visible

Paper Containing Fused Thermoplastic Material Distributed in a Regular Pattern

H.M.G. Williams; U.S. Patent 3,880,706; April 29, 1975 provides a method of making security paper which comprises incorporating in the body of the paper thermoplastic material and raising the temperature of the thermoplastic material to a level at which the material fuses.

When the material fuses, it becomes amalgamated with the fibers adjacent to it, that is to say it impregnates these fibers, thus becoming inseparable from the paper. Delamination is thus made virtually impossible. Also, the presence of the thermoplastic material can, depending upon the particular properties of the material and of the body of the paper, be detected by examination under transmitted light, by direct light or by feel.

Various thermoplastic materials may be employed. Suitable materials with appropriate fusion temperatures below that at which the paper is adversely affected include polyamides, polyvinylchloride and low melt polyethylene. Preferably the material fuses below 130°C.

The thermoplastic material prior to incorporation in the paper may be in various forms and it may be incorporated in various ways. The most preferred form of the thermoplastics material is a polymeric net which has been so stretched that the filaments between the nodes of the net have their molecules orientated. Such a net may be made by extruding a polymeric film, embossing it in such a manner with an embossed surface that when the embossed film is subsequently stretched in the cross and machine direction the film structure expands to produce a net-like structure in which the cross-over points or bosses are linked to other contiguous bosses by thinner strands of polymeric material.

Where the thermoplastic material is in continuous form prior to incorporation in the paper, it may be incorporated during the actual deposition of paper stuff to form the paper web. For example, the material may be fed on the paper-forming surface of a papermaking machine (e.g., onto the cylinder of a cylinder paper machine or onto the wire of a Fourdrinier machine) in such manner that the material contacts the cylinder at a point at which a proportion, preferably about half, of the substance of the desired paper web has already been deposited on the surface.

Thus the remainder of the substance of the sheet will be deposited over the material, drainage taking place through the material. In this way the material, e.g., a polymeric net, may be completely incorporated into and enclosed within the freshly-formed sheet of paper. The resulting sheet normally can be processed on conventional vacuum boxes and drying systems and the waterleaf structure can normally be sized in the conventional manner.

Where the thermoplastic material is in continuous form such as a net, it is preferably dispensed from a container in which it is arranged concertina fashion so that a further strip can be joined to the end of the strip being fed into the paper forming machine before it is exhausted so that papermaking can proceed continuously. Strips can be joined, for example, by water-insoluble cements or by heat fusion. The process is described with reference to Figure 3.1.

FIGURE 3.1: PAPER CONTAINING FUSED THERMOPLASTIC MATERIAL DISTRIBUTED IN A REGULAR PATTERN

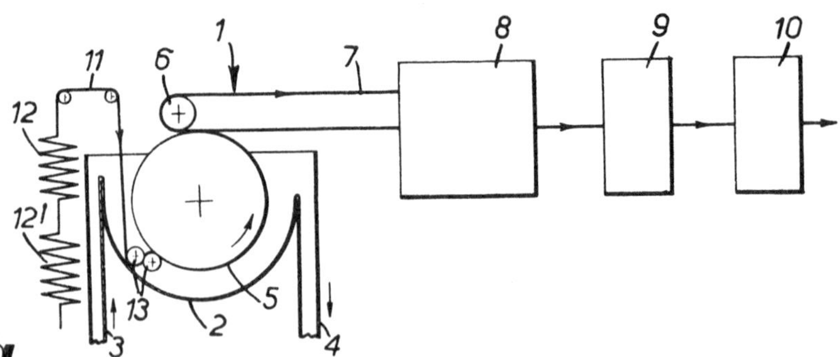

a.

Single Cylinder Papermaking Plant

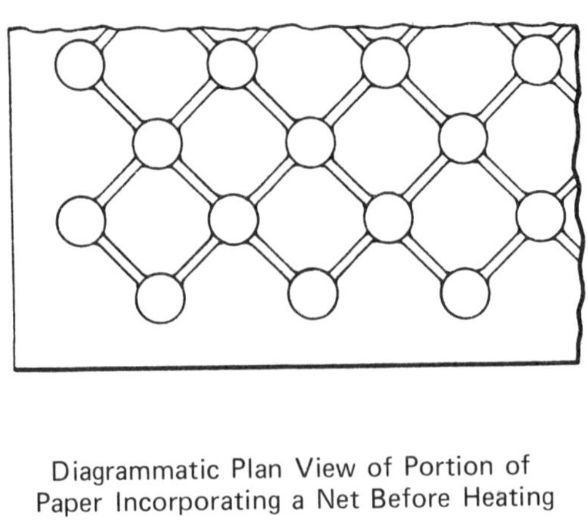

b.

Diagrammatic Plan View of Portion of Paper Incorporating a Net Before Heating

(continued)

FIGURE 3.1: (continued)

c.

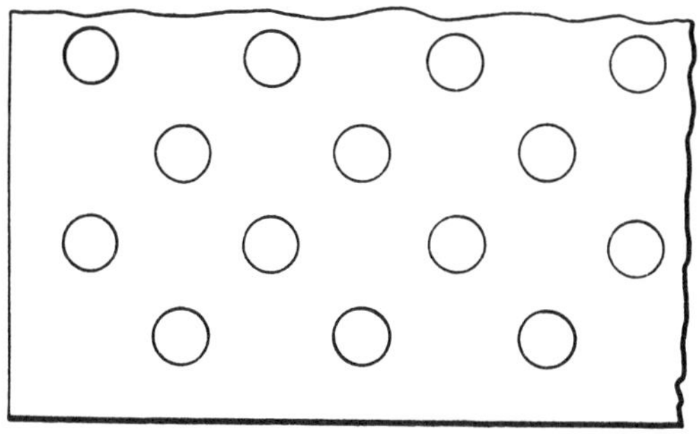

View of (b) After Heating

d.

Cross Section of (c) on an Enlarged Scale

Source: U.S. Patent 3,880,706

The cylinder papermaking machine **1** has a vat **2** with an inlet **3** for paper stock and an overflow outlet **4**. Paper is formed on the periphery of a screen cylinder **5** from which it is removed by a couch roll **6** and carried away on a felt **7** to a drying zone **8** in which, for example, the paper may be dried between felts which pass over steam-heated drums. After drying the paper is passed to a sizing facility **9** and then passes to a heating zone **10**.

A strip of thermoplastic net is incorporated in the paper as it is being formed on the cylinder **5**. This is achieved by feeding the net **11** in continuous strip form from a package **12** into the vat **2** and onto the surface of the cylinder **5** at a point **12** where some paper-forming fibers have already been deposited on the cylinder. Adjustable guide rollers **13** direct the net web **11** onto the cylinder at the selected point. The package **12** contains the net strip in concertina folded form with a tail **15** which is joined by an adhesive or by welding to the head of a reserve package **12'**. As soon as the package **12** is exhausted the package **12'** begins to supply the net and a further package will be attached to the tail of the package **12'**.

Example 1: A paper was made on the plant shown in Figure 3.1a incorporating a net prepared in the manner described above and in British Patent No. 914,489. The net was of high density polythene colored red by the inclusion of pigment during extrusion, and was so embossed that it contained an average of 97 bosses per square inch, and 30 yards of 38 inches wide net weighed 1.1 lb. The net was cut into tapes 5 cm wide and fed into the vat **2** of the cylinder forming machine **1** on which an 80 gsm (grams per square meter) beaten cotton paper was being formed.

The angle and point of contact of the net band with the wire was adjusted by the guides **13**, such that the amount of stock deposited on the wire before the net contacted the cylinder was approximately 40 gsm. It was observed that under these conditions sufficient drainage occurred through the net to result in a total deposition of 80 gsm of paper fibers, with the paper formed through and around the net.

The web was couched on to the felt **7**, dried over drying cylinders in the normal fashion in the drying zone **8**, and then tub sized in the sizing zone **9**. A diagrammatic plan view of a portion of the paper at this stage is shown in Figure 3.1b.

The web was then passed through the heating zone **10** which contained infrared heaters by which the temperature of the paper was raised above the fusing temperature of the included net. After calendering the paper was examined, and it was found that on tearing no trace of the net structure could be detected. The red polythene dots which represented the residual bosses were plainly visible inside the paper by both transmitted light and reflected light. The presence of the security device, i.e., the dots, could be detected by feel.

A diagrammatic plan view of the finished paper is shown in Figure 3.1c and a cross section to an enlarged scale is shown in Figure 3.1d. It will be seen that the filaments between the nodes of the thermoplastic net have, under the influence of heat and because of the orientation of their molecules, broken and have drawn back into the nodes to leave a pattern of dots. These dots are faintly visible in direct light and are plainly visible in transmitted light. It will be seen from Figure 3.1d that there is a slight elevation of both surfaces of the paper adjacent the nodes and these can be detected by touch. It will also be seen that the thermoplastic material has mingled with the adjacent fibers of the paper.

The fused thermoplastic material thus gives the paper features which are very difficult for the forger to reproduce but whose presence can readily be detected; at the same time the material makes delamination of the paper extremely difficult. Various effects can be produced by appropriate treatment of the thermoplastic material prior to incorporation.

Example 2: Thermoplastic Identifying Agent Treated with Metallizing Agent — A 300 denier polyamide terpolymer containing 66 nylon, 12 nylon with a melting point of 120°C was metallized by the vacuum deposition of aluminum. The monofil was then fed onto the cylinder papermaking machine shown in Figure 3.1a in such a manner that the thread contacted the cylinder after 40 gsm of paper fibers had been deposited. In the residual rotation of the cylinder a further 40 gsm of paper fibers were deposited. The paper web was couched and dried by passage over steam heated rollers in the zone **8**. During this latter operation the polyamide thread fused so that the separate heating zone **10** could be omitted.

Inspection of the paper by transmitted light showed the presence of a metallized thread, but it was impossible to split the paper on this line, or to withdraw the thread.

Example 3: Thermoplastic Identifying Agent Treated with Printing Agent — A polythene monofil with a melting range starting at 120°C was printed with nitrocellulose inks, by passing through a series of rollers each of which was synchronized to deposit a 2 mm long ink mark at 5 mm spacing onto the thread. In this manner a series of colors were placed on the monofil.

The monofil was incorporated into a paper in the manner described in Example 2. On inspection the dried paper showed both by refracted and reflected light a series of spaced colored dots inside the paper.

Example 4: Thermoplastic Identifying Agent Treated with Opacifier — A paper was prepared as in Example 3 in which the monofil was pigmented with titanium dioxide to produce such an opacity that in the dried paper the thread was not detectable but only the colored dots were visible.

Anticounterfeit Betting Ticket

K.P. Rowsam and D.W.M. Yost; U.S. Patent 3,677,887; July 18, 1972; assigned to The Sorg Paper Company describe the manufacture of an anticounterfeit ticket formed from a paper sheet which permits the ink used to print the indicia on the ticket at the time it is issued to penetrate completely through the sheet so that the indicia will be readily visible from either side. The paper is manufactured utilizing conventional papermaking equipment, usually a Fourdrinier papermaking machine.

Unlike conventional tickets, the sheet stock in accordance with the process has a basis weight of from about 45 pounds per ream to about 102 pounds per ream at a thickness of from 0.007 to 0.012 inch, which results in a density of from about 6.5 to about 8.5.

The furnish for the base stock comprises from about 40 to about 75% of a high alpha saturating pulp which contains 92% or more alpha cellulose. The remainder of the furnish may comprise a chemical wood pulp. Semibleached kraft is preferred, particularly when blended with a mercerized soft wood sulfate pulp. The pulp may be treated and blended in conventional fashion and delivered to the headbox of the papermaking machine.

In order to develop directional stiffness in the stock, it is desirable to have no pressure in the headbox, and to maintain as low a head as possible, preferably 4 inches at machine speeds of approximately 300 feet per minute. In the papermaking machine, the furnish will be felted in the usual manner, the formed web passed through the conventional press and driers, whereupon the web is delivered to the size press for the sizing treatment and subsequent finishing. The finished sheet will preferably have a caliper of 0.0085 to 0.0095 inch, although additional latitude is permissible and the caliper of the sheet may vary between 0.007 and 0.012 inch.

The sizing of the sheet is also important, the primary consideration being the utilization of a size which is not a film former but rather will strike into the sheet and lock the fibers, yet the size must not block the interstices in the sheet so as to prevent strike-through of the mineral spirit ink. Preferably, the size is a combination of three materials, namely, starch, thermosetting resin and wax, which will be applied by a shower on the wet side of the papermaking machine, i.e., in advance of entering the sizing press.

The major constituent (67 to 83%) of the size is a starch characterized by a high solids content at low viscosity. The starch constituent of the size should have minimum viscosity of 35 seconds as measured with a Dudley pipette at 21% solids at 140°F and a maximum viscosity of 55 seconds under the foregoing conditions. An oxidized potato starch may be employed, as may be any other type of oxidized converted starch. Cornstarches may also be employed, such as an acid-thinned hydroxyethylated cornstarch. It may be noted, however, that while cornstarch readily lends itself to crosslinkage with the resin constituent of the size, a size composed of all cornstarch is not desirable since the sheet becomes too stiff and tends to be brittle. A preferred starch constituent comprises a blend of equal parts by weight of an oxided potato starch and hydroxyethylated cornstarch.

The resin should be of the thermosetting type which is a formaldehyde condensation resin, such as urea-formaldehyde or melamine-formaldehyde. The latter is preferred in that it has a greater pot life than the urea types. The resin constituent should be present in an amount of from about 15 to 25% by weight of the size.

Numerous waxes may be employed, inclusive of paraffin, microcrystalline, stearic and natural, the essential consideration being the use of a wax emulsion which is stable between the range of pH 6 to 8 and has a melting point above the 140° to 150°F temperatures encountered during application. A preferred wax is a 44% solids emulsion consisting of 80% wax and 20% rosin. The wax constituent preferably comprises about 5% by weight of the size although it may be varied from about 2 to 8% on a dry weight basis.

A preferred sizing composition contains 73% starch, 22% thermosetting resin and 5% wax, the total solids content being approximately 15% by weight. In terms of a size pickup, the sheet should contain from 10 to 20% size based on the total weight of the sized sheet. Using the preferred sizing composition set forth above, a 14.5% size pickup is preferred.

Following the application of the size, the stock is fed into another drier section of the papermaking machine where it is dried to a moisture content of about 4%. In this connection, the drying of the sheet to a low moisture content assists in insuring maximum crosslinkage between the thermosetting resin and the starch. Where the thermosetting resin comprises a melamine-formaldehyde resin, it has been found that the resin is 90% cured as it is removed from the paper machine and reaches 99% cure after aging in rolled condition for approximately 21 days.

In finishing the sheet stock it is preferred to calender it through a single nip to provide a minimum surface finish which will retard the penetration of the printing ink. Stiffness along with penetrability by the initial spirit ink are prime requisites of the sheet. A dry stiffness of 900 mg is preferred, as measured on a Gurley Stiffness Tester, the stock being measured in an atmosphere of 50% relative humidity at a temperature of 72°F. While a wet stiffness of at least 500 mg is preferred, the stock being measured after being conditioned for 24 hours at 90% relative humidity and 80°F, stiffnesses as low as 350 to 375 mg are operative although performance is marginal in the ticket dispensing machines.

It should also be explained that the preprinting of, for example, a track ticket is generally performed on a Flexographic press using color pigmented ink in a latex binder. Such inks dry by evaporation and hence do not rely wholly on absorption.

As the ticket is sold and dispensed, the dispensing machine will print additional pertinent information on the ticket, such as the amount of the bet, the number of the horse or dog, the number and date of the race and the track code (it will be understood that the indicia will vary).

It is this latter ink which penetrates completely through the sheet and is readily visible from the reverse side in reverse image. Such inks normally comprise a hydrocarbon vehicle, such as mineral spirits, and a combination of dyes and pigments which are insoluble in the normal bleaching agents.

The process thus provides a betting ticket which is readily receptive to conventional mineral spirit printing inks and will permit such inks to strike completely through the sheet so as to be visible on the reverse side without undue bleeding or distortion of the printed indicia, the sheet at the same time exhibiting a stiffness preferably within the range of 450 to 550 mg at a relative humidity of as much as 90% at 80°F.

SHADOWMARKS

Epoxy Resin Application Followed by Starch Size Before Resin Cures

B.D. Skofronick; U.S. Patent 3,441,427; April 29, 1969; assigned to Customark Corporation describes a method in which opaque areas may be imparted either in a selected design or entirely throughout the article by few manipulative steps. The surface of a sheet of paper is variously treated to form thereon opaque areas that will be serviceable in generally the same way as the original paper.

The process may be summarized in the following steps:

(1) Impart a nonaqueous liquid composition containing thermosetting resin and curing agent to surface of a finished sheet of paper;

(2) Contact the resin and curing agent with a sizing solution before the resin cures into a hardened form; and

(3) Cure the thermosetting resin with the contacted sizing solution to render the configuration less translucent than the balance of the paper.

The first step in the process uses a similar liquid composition containing thermosetting resin and a curing agent (impregnating materials) as disclosed in U.S. Patent 3,140,959. The liquid composition includes organic solvent which, after removal by evaporation, would render the paper translucent in the area of the hard chemical watermark. The liquid composition may be applied to the surface sheet of paper by various means, such as printing or brushing the mark through a stencil. The area is manipulated by another step wherein a sizing solution contacts the area before the thermosetting resin in the liquid composition has hardened into what would be the chemically translucent area according to U.S. Patent

3,140,959. After the liquid composition with the impregnating materials contacts a sheet of paper the solvent in the liquid composition begins to evaporate. It is preferred that the step of contacting the area of the impregnating materials with the sizing solution occurs prior to evaporation of all the solvent. The important feature is that contact with the sizing solution should occur before the synthetic thermosetting resin cures into a hardened film. Various sizing solutions are commonly used in the paper, and the most popular are the starch solutions, such as the chlorinated and ethylated starches.

Following this step, the resin area has not yet attained its opaque state, or a condition in which such area is less translucent than the original condition of the paper. The last step requires that the resin area, contacted with a sizing solution, is cured into a hard flow. This curing step is preferably by heating, and this step results in the area becoming opaque (less translucent) than the remaining portion of the paper. Heating is preferred and not critical because this merely enhances or accelerates the rate of cure. Particularly successful opaque areas have been obtained by employing an epoxy resin and appropriate curing agent as the impregnating material in the liquid compositions.

Example 1: Opaque Shadow Mark — A liquid composition, containing an epoxy impregnating compound, is prepared from the following ingredients and amounts:

Ingredients	Parts by Weight
Epoxy resin (Dow Der 332)	5
Epoxy hardener (Dow experimental hardener X-2654.4)	1
Methyl Carbitol (diethylene glycol monomethyl ether)	4.5

A small metered amount of liquid composition is deposited on the surface of a foot square piece of ½" thick plate glass. The composition is spread uniformly over the surface with a hand roll, consisting of an engraved roll bearing 75 depressions per inch and fitted with a handle. A die or hand stamp of rubberlike material having inscribed on the face thereof a typical watermark design is first pressed against the wet film of the composition on the glass and then stamped on the paper sheet. Before the epoxy resin is cured by the curing agent into a hard film, the sheet of paper is totally immersed in a bath containing 8% chlorinated starch in water, the bath maintained at 80° to 150°F. The sheet of paper is removed and subjected to heating by hot air until the paper is dry.

Example 2:

Ingredients	Parts by Weight
Epoxy resin (Dow Der 332)	5
Pyromellitic dianhydride	1
Methyl Carbitol	4.5

The above liquid composition, with epoxy resin impregnating compound, is manipulated in the same way described for Example 1 to obtain a shadow chemical watermark of desirable properties.

The pyromellitic dianhydride has been found to be a preferred curing agent or hardener for the epoxy resin, since likelihood of possible discoloration of a chemical watermark is even further reduced. The pyromellitic dianhydride is preferred curing agent, not only for the shadow chemical watermarks but for the translucent chemical watermarks.

Example 3: Opaque Sheet —

Ingredients	Parts by Weight
Epoxy resin, Epon 812	6
Epon curing agent H-3	3
Methyl Carbitol	9

The above ingredients are mixed and a portion of a few ml is deposited at the top of a sheet of paper. A gravure roller of a width wider than the paper is pulled through the deposited liquid composition causing the composition to be laid down on a sheet of dry or finished paper. Within 3 minutes the entire sheet is immersed in a bath containing 8% chlorinated starch and held at 80° to about 150°F. The excess starch solution is removed and the sheet is exposed to a stream of hot air until the sheet is dried and becomes opaque relative to its original condition.

Resin Treatment Followed by Water Treatment

This method of *B.D. Skofronick; U.S. Patent 3,486,923; December 30, 1969; assigned to Customark Corporation* provides that opaque areas are made on paper by applying a liquid composition containing one or more resins to the surface of the paper so that the resin is laid down as a wet film and impregnates the paper fibers or the paper web. The composition is in a liquid form to facilitate application, and has a viscosity which is sufficiently low to allow the desired impregnation.

When opacifying entire sheets of paper, it is only required that the viscosity be not so excessively low so that the impregnating composition would pass through the paper. When making opaque shadowmarks, the viscosity is sufficiently high to retain the design which has been applied to the surface of the paper. In other words, the viscosity is low enough to impregnate, but not low enough to distort the preselected design which has been applied to the surface.

The liquid resin film or layers which have been applied to the surface of the paper are then rendered more opaque than the original paper by contacting the resin with water. This step leads to attaining an opaque area which is further raised relative to the original surface of the paper.

Such results would not occur without the water treatment. In fact, without water treatment the area would become translucent relative to the original condition of light transmission, and the area would not be essentially raised relative to the original surface of the paper.

In some instances, a single water treatment step will lead to forming an opaque area but, in many instances, this will not lead to a desirably formed opacity. A plurality of successive water treatment and intervening drying steps can lead to the development of an opaque area, or to the formation of better defined opacity.

The water treatment step may be performed in a variety of ways such as spraying, fogging, dipping, providing a high humidity environment or by other means. A successful step provides for dipping the sheet of paper in a body of water, then nipping the paper between rollers to remove excess water, and then air drying or drying at moderate temperature levels. When using a plurality of successive wetting and drying steps, the sheet may be dipped in water and nipped between rollers, dried in an oven at temperatures of about 105°C or over a hot felted dryer held at 78° to 80°C, dipped again, nipped, dried again and so on.

The term "resins" is intended to cover materials which can be applied as a wet film which can impregnate paper fibers or the paper web, and which materials can be contacted with one or more water treatments to increase the opacity of the resin-treated area. The resin may be of the class of a natural resin, a substituted resin and a modified resin. The resin may be a thermosetting resin of the class of an epoxy, a polyester, an alkoxy substituted melamine, a urea-formaldehyde and a melamine-formaldehyde, used with or without a curing agent.

The resin material may be of the class of Tripropionin, N-methyl 2-pyrrolidone and diethylene glycol monomethyl ether, a polypropylene glycol having a molecular weight of about 400, an alkyl ketene dimer, a mixture of an epoxy and an acrylic resin, a mixture of ethylene glycol and acetylenic glycol, a polyvinyl chloride, a dicarboxylic anhydride plasticizer, an organo-silicone copolymer or a saturated polyester.

Example 1: Variety of Resins — (a) Eight impregnating liquid compositions, designated A through H, respectively, are prepared as follows:

Liquid Composition	Ingredients
A	Monomer X 970 (Rohm & Haas Co.)
B	Poly Glycol P-400 (polypropylene glycol, MW 400)
C	Five parts Marvinol (polyvinyl chloride); ten parts Tripropionin
D	Five parts Epocryl E 11 (epoxy and acrylic combination); ten parts methyl Carbitol
E	Surfynol 485 (reaction product of ethylene oxide and acetylenic glycol)
F	Pegosperse 400 (stearate and glycol combination, MW 400)
G	Methyl Carbitol
H	Tripropionin

(b) Strips of various paper weight are provided in dimensions of 4¼" x 11". The strips are dipped in the various compositions. The strips are then nipped between a brass roller and a rubber-covered roller, and then passed over a hot felt drum dryer for 60 seconds at 79°C. The strips were then treated with water. The strips are dipped in the water, nipped between the described rollers and then dried by being passed twice over the felted drum for a total drying time of 120 seconds. This series of dipping, nipping and drying is followed for each additional water treatment. Opacity measurements were made on control strips and on the water-treated strips. The opacity measurements represent the average of three readings on spaced portions of each strip, and they are collected with a Bausch & Lomb opacimeter.

(c) The following table shows the opacity measurements of the above eight liquid impregnating compositions which were applied, all of which were used on 9 lb paper. The data shows that for compositions A through D, inclusive, the opacities have been uniformly increased relative to the blank opacity by the fifth water treatment. For liquid impregnating compositions E through H, inclusive, the opacities have been all increased relative to the blank by the sixth water treatment. Such increase, in general, has continued with successive water treatments except for compositions E and H which show some slight reductions in opacity by the twelfth treatment.

	Liquid compositions							
	A	B	C	D	E	F	G	H
Blank	54.3	54.3	54.3	54.3	54.3	53.8	54.9	54.2
1st water treatment	42.4	32.8	29.8	42.9	57.6	42.7	55.8	28.6
2nd water treatment	55.1	43.5	37.3	48.6				
3rd water treatment	55.5	50.2	49.2	51.9	58.3	49.8	56.6	47.4
4th water treatment	52.4	49.3	61.1	54.7				
5th water treatment	55.5	54.5	62.7	55.0				
6th water treatment					57.6	57.1	55.9	64.0
9th water treatment					59.1	58.6	56.0	61.5
12th water treatment					58.7	64.0	56.8	56.8

Example 2: (a) Thermosetting Resin plus Curing Agent — A liquid composition is prepared from the following ingredients:

Ingredients	Parts by Weight
Hexamethoxymethylmelamine, Cymel 301	10
EPI Cure 8771, a Lewis acid	3
Methyl Carbitol	2

Paper strips of 9 lb weight are used and procedures of Example 1 are followed with the above liquid impregnating composition to obtain a replica number of opaque areas after one and after two water treatments. The results show that a second water treatment was sufficient to opacify the area, while the first water treatment was insufficient for this purpose.

(b) Thermosetting Resin Alone — A liquid composition is prepared from the following ingredients:

Ingredients	Parts by Weight
Hexamethoxymethylmelamine, Cymel 301	7.5
Methyl Carbitol	2.5

The above liquid composition is similar to (a) except that no curing agent is included. The same paper and process steps are followed. In several instances, one water treatment was sufficient to opacify the paper, and two water treatments were adequate to opacify the papers in all instances.

The examples have illustrated the method of opacifying entire strips of paper. It should be understood that the same conclusions will operate relative to the preparation of shadowmarks (see U.S. Patent 3,441,427).

Resin Application, Water Treatment, Solvent Treatment

In a further modification *B.D. Skofronick; U.S. Patent 3,443,979; May 13, 1969; assigned to Customark Corporation* provides that the opaque areas are developed in an improved manner by contacting deposited resin on the paper with an organic solvent after the water treatment step, preferably after the paper has been dried following the water treatment step.

The process, which uses a solvent in one or more solvent treatment steps, leads to a recognizable and measurable opacity. The perceptible opacity is recognized by visual observation and the measurable opacity is one in which the opaque area is raised at least about 5% relative to the original thickness of the paper and in which the opacity is increased at least 1% relative to the light transmission of the original paper. While the foregoing measurable indices are often attained with one water treatment step alone and nearly always with two or more water treatment steps, the development of the opacity is enhanced by employing one or more solvent treatment steps following such water treatment steps.

It is an advantage that the use of the solvent treatment steps can lead to the attainment of an opaque area of desired quality with an overall fewer number of treatment steps. The use of the solvent treatment step will lead to an opaque area having a quality which would otherwise be obtained only with additional water treatment steps alone.

The drying step should remove substantially all of the water, down to about 2% in the paper. In general, sheets of paper have been effectively dried by passing them over heated, felt-covered drum driers having a surface temperature of about 78° to 80°C. The paper sheets are effectively dried when they are brought into contact with such heated drums for short periods of time of about 60 to 120 seconds. Likewise, hot air or heating ovens may be employed for correspondingly brief periods of time to remove substantially all of the water.

Many types of organic solvents are operable. These include the aliphatic hydrocarbons, the halogenated hydrocarbons, alkanols, the ketones, esters of monocarboxylic acid and medium length carbon alcohols, the hydrocarbon ethers, substituted and unsubstituted cyclic hydrocarbons which may be saturated or unsaturated, polyhydroxyl alcohols, esters of alcohols and dicarboxylic acids, and alkyl anhydrides.

Here too the term "resins" is intended to cover materials which can be applied as a wet film which can impregnate paper fibers on the paper web, and which materials may be contacted with one or more water treatments to opacify the paper. Such paper impregnating materials may include natural and synthetic resins, plasticizers, some organic solvents, esters and other materials.

The natural resin may be a rosin, a substituted rosin (hydrogenated methyl ester rosin), a modified rosin (a partially decarboxylated rosin) or a substituted and modified triethylene glycol ester of hydrogenated rosin. The resin may be an ester of a polyol and a saturated diacid or an unsaturated diacid, a dialkyl sulfosuccinate, sucrose acetate isobutyrate, cyclohexanone or a condensation product of cyclohexanone and methyl cyclohexanone. The resin may be a monomeric plasticizer used as a diluent for an aryl sulfonamide, a polymeric plasticizer, a polyalkylene glycol (propylene glycol), ABS or an alkyl ketene dimer.

The resin of the liquid composition may include a thermosetting resin of the class of an epoxy, a polyester, a urea-formaldehyde, a melamine-formaldehyde and hexamethoxymethylmelamine, with or without a curing agent.

Example 1: Shadowmark — A liquid composition is prepared by mixing the following ingredients in the indicated amounts.

Ingredients	Parts by Weight
Epoxy resin (Iso Chem Res Trans Flex)	10
Epoxy curing agent (EPI Cure 8771)	3
Methyl Carbitol (diethylene glycol monomethyl ether)	10

The ingredients are mixed to attain their liquid composition, and a small metered amount of the liquid composition is deposited on the surface of a foot square piece of ½-inch thick plate glass.

The composition is spread uniformly over the surface with a hand roll consisting of an engraved roll bearing 75 depressions per inch and fitted with a handle. A die or hand stamp of rubber-like material having inscribed on the face of a typical watermark design is first pressed against the wet film of the composition on the glass and then stamped on a paper sheet. About 3 minutes later, the sheet of paper is totally immersed in a bath of water maintained at room temperature. The sheet of paper is withdrawn, the excess water is removed by a nipping operation, and the paper is subjected to heating by contact with a hot felted dryer maintained at 80°C for 60 to 120 seconds.

Following a single water treatment, the opacity of the treated area may actually be decreased as compared to the original light transmission of the paper. The paper, however, is dipped in acetone following the drying and then again dried with hot air as before. The opacity of the area, following such solvent treatment, is increased substantially relative to the opacity of the area after the water treatment.

Another sheet of paper has the same liquid impregnating composition applied in the form of a selected design. The sheet is dried by passing over a hot felted dryer maintained at 80°C for 60 seconds. The sheet is then dipped in water and nipped between a brass roller and a rubber-covered roller to remove the excess water. The wet sheet is then passed over the hot felted dryer twice, each pass taking 60 seconds to execute. The sheet is then dipped in water a second time and dried as in the previous sequence of dipping and drying. Following these two water treatment and drying steps, the sheet is dipped in acetone, nipped between the brass and rubber rollers, and then passed twice over the hot felted dryer held at 80°C, each pass requiring 60 seconds.

Example 2: Opaque Paper — (a) The liquid composition of Example 1 is used to cover entire paper strips. A plurality of paper strips which are 4¼" x 11" are immersed in the above liquid impregnating composition, then nipped between brass and rubber-covered rollers, and dried by passing over a hot felt-covered drum dryer at 70°C for 60 seconds. The dried strips are then dipped in water, nipped as in the resin treatment step and dried by being passed twice over the drum dryer for a total of 120 seconds of drying time. After the first water treatment, the paper is again dipped in water, nipped and dried. The paper strip is then dipped in acetone and dried by passing twice over the drum dryer for a total drying time of 120 seconds.

Following one and two water treatments and drying steps, the strips are treated with a variety of solvents and with a variety of solvent mixtures. The representative organic solvents employed are listed on the following page together with the corresponding symbols.

Organic Solvent	Symbol
Acetic anhydride	AcAn
Acetone	A
Amyl acetate	AmAc
Ethyl alcohol	EA
Ethylene glycol	EG
Carbon tetrachloride	CT
Heptane	H
Methyl Carbitol	MC
Toluene	T
Trichloroethane	TCE
Ethylene	X

The paper used in such tests is 9 lb paper, 25% cotton (500 sheets, 17" x 22" per sheet).

The following table shows that a liquid composition containing an epoxy resin and a chemical curing agent for the resin leads to an opacity increase in each instance after the first solvent treatment relative to the first water treatment. After a solvent treatment following two water treatments, the opacity increases, in each instance, over the light transmission of the original paper and over the measured opacity following the second water treatment.

Solvent	Blank	First Water Treatment	Solvent Treatment	Percent Change	Second Water Treatment	Solvent Treatment	Percent Change
A	54.5	43.8	66.5	34	50.7	68.3	26
CT	54.9	45.5	61.6	26	56.7	62.3	9
H	53.8	42.1	46.9	10	51.3	55.7	8
X	55.1	43.0	50.4	15	55.8	65.2	14
TCE	53.1	41.0	46.8	12	50.9	60.0	15
MC	52.6	35.8	49.3	27	45.7	61.7	26
T	54.0	36.5	36.0	-	46.6	62.2	25
Am Ac	52.1	37.6	56.1	33	49.5	61.9	20
EG	54.5	36.9	61.7	40	46.3	61.2	24
1A:1H_2O	52.3	43.3	50.7	15	54.8	59.2	7
1EA:1H_2O	53.2	44.8	53.3	16	51.8	58.0	11
1CT:1H_2O*	54.3	42.2	50.0	16	52.1	58.0	10
1A:1CT	52.0	55.4	63.0	12	61.8	69.8	12
1A:3MC	53.1	35.8	59.0	39	46.6	65.2	29
1A:1MC	52.4	37.4	53.7	30	62.2	63.7	2
3A:1MC	54.9	42.2	49.5	15	61.6	69.3	11

*Immiscible mixture.

(b) A liquid composition containing an epoxy resin but no curing agent leads to an increased opacity following organic solvent treatments, in every instance, over the corresponding first and second water treatment.

Example 3: Variety of Resins — A variety of liquid impregnating compositions were prepared, and paper strips were dipped in each liquid composition.

Impregnating Material	Solvent
Staybelite ester No. 3	-
5 parts Dresenol 142	5 parts methyl Carbitol
5 parts Santolite MS 80% solution	5 parts methyl Carbitol
Monoplex S-73	-
5 parts Paraplex G-40	5 parts triacetin (Armour)
5 parts Eastman Polymer Plasticizer NP-10	5 parts methyl Carbitol

(continued)

Impregnating Material	Solvent
5 parts Santolite MHP	5 parts methyl Carbitol
Monomer X-970 (butylene dimethacrylate)	-
Dow Polyglycol P-400	-
5 parts acrylonitrile-butadiene-styrene	5 parts MEK plus 20 parts triacetin
5 parts polyvinyl chloride	10 parts Kodaflex Tripropionin

Staybelite ester No. 3 — A triethylene glycol ester of a hydrogenated rosin
Dresenol 142 — A partially decarboxylated pale rosin
Santolites — Aryl sulfonamide formaldehyde resins
Paraplex — Linear polyesters of aliphatic dicarboxylic acids and glycols
Polyglycol 400 — Polypropylene glycol plasticizer, MW 400
Monoplex-73 — A monomeric epoxide ester
Morester X-973 — A saturated polyester
Eastman polymer plasticizer — A polymer derived from neopentyl glycol

The data indicated that a second successive acetone treatment even further increases the opacity of the strips, although the increase is not so dramatic as the increase of the first acetone treatment over the water treatment or treatments.

ABSORBENT PAPERS

CREPED TISSUE

Stiff Crosslinked Fibers plus Refined Papermaking Fibers

The creping of paper on a Yankee drum surface is well known. It involves crowding the tissue paper sheet in the longitudinal direction while it is adhered to the drum to cause disruption of fiber-to-fiber bonds, that is, to limit fiber-to-fiber bonding. The creping action may take place when the sheet is well dried on the Yankee or while it is in the wet, somewhat plastic state.

L.J. Bernardin; U.S. Patent 3,455,778; July 15, 1969; assigned to Kimberly-Clark Corporation incorporates in a papermaking furnish with the usual papermaking pulp fibers a substantial quantity of significantly stiff crosslinked wood pulp cellulosic fibers of papermaking length.

The characteristics of the stiff fibers include: a substantial lack of interfiber bonding capacity for each other and for the papermaking fibers in the wet and dry state, naturally occurring kinks and curls of the original wood pulp fibers set in the fibers by crosslinking, a substantial fiber rigidity but subject to resilient compression when wet, and a distinct capacity for being wetted by aqueous fluids.

The furnish is a uniform blend of the fibers which are interfelted together in a sheet by waterlaying. The conventional steps of dewatering and wet pressing follow and commonly the sheet is presented on a Yankee dryer surface to a creping blade.

The sheet formation, water content reduction and wet pressing provide a wet sheet in which the interfelted fibers are pressed together and lie substantially parallel to the same plane. The usual cellulosic papermaking fibers, have an inherent fiber-to-fiber bonding capacity. This capacity is very pronounced when the sheet is dry to the touch and is commonly termed hydrogen bonding. In the somewhat wet plastic state a significant, though lesser fiber-to-fiber bonding capacity, exists between papermaking fibers. The stiff crosslinked fibers lacking in significant bonding capacity but interfelted with the papermaking fibers become retained by the latter. Additionally, the kink and curl of the crosslinked fibers and the resilient compressibility of the fibers apparently permits a significant force to be exerted upon the papermaking fibers serving as retainers.

Then when the sheet is dried on the Yankee to effect at least some drying, fiber-to-fiber bonds develop between the papermaking fibers. The web is then presented to the creping blade, and bonds between the cellulosic papermaking fibers are disrupted by creping action. The more dry the sheet as it is presented to the blade the more it adheres to the Yankee dryer surface and the greater is the disruptive effect of the blade. Additionally, usually the greater the adhesion the more of the hill and valley effect common to crepe sheets is attained.

Release of the bonds between the cellulosic papermaking fibers by creping action is facilitated by the stiffened fibers. These fibers, having been held under compression in the plane sheet, urge the cellulosic retaining fibers apart, providing for very open pores between fibers. More noticeable visually is the development of a nap on the sheet surface. This is occasioned by the stiff fibers springing up from the web surface as they are released. Importantly, these stiff fibers project in all directions within and from the sheet surface including bridging of hills and valleys of the crepe sheet.

The resulting product of the procedure as described surprisingly exhibits a much more square absorbency characteristic than do crepe sheets of only the usual cellulosic papermaking fibers.

With known tissues having a distinct crepe, that is, a ratio 1.1:1 and above, the wicking rate for fluids is much greater along the hills and valleys and the absorbency pattern for a drop of fluid, oil or water, is generally elliptical with the long axis in the valley direction. As the crepe ratio increases the disparity between the valley direction and the cross-valley or longitudinal direction of the sheet increases.

In the crepe sheet of this process the absorption characteristic for a drop of fluid at low crepe ratios is substantially circular and very nearly so even at high crepe ratios. Preferably, the crepe ratio is between about 1.2:1 and 3:1. In some instances a nearly circular absorbency characteristic may be obtained with conventional fibers by drawing out a large proportion of the crepe and calendering but this is at the expense of transverse wicking rate. In this case transverse wicking rate is maintained and the circular pattern of absorbency is evidence of the increased total wicking rate of the sheet of this process.

Example: (a) Stiffened wood pulp fibers were prepared by first forming a mat of bleached southern pine pulp; this mat was then saturated with an aqueous solution of dimethylolurea in which the dimethylol was present to the extent of about 12½% by weight. The solution included also as catalyst ammonium chloride to the extent of about 2% by weight of the dimethylolurea. The pulp mat, after saturation with the solution, was pressed to a 200% wet pickup and air dried. This sheet was then defibered in a hammer mill to form a fluff consisting essentially of individual fibers. The fluff was cured at about 130° to 135°C in order to effect crosslinking.

(b) A usual papermaking pulp for tissue purposes, that is, a 50-50 blend by weight of bleached softwood sulfite pulp and a spruce kraft pulp were made up as a papermaking furnish. For this purpose the pulp blend was refined to a Canadian Standard Freeness of about 550. Additionally, the pulp combination included 1% by weight based on the pulp weight of a wet strength resin.

(c) About 30% by weight of the stiffened fiber in aqueous suspension and produced as described above was added to 70% by weight (air-dry fiber basis) of the pulp blend to form the papermaking furnish.

(d) A tissue sheet was formed on the papermaking machine in customary manner, by passing the furnish at about 0.2% consistency to a wire of a Fourdrinier machine. From the Fourdrinier wire the formed sheet moved on a felt and then to a highly polished surface of a Yankee dryer against which it was laid by means of a press roll, the paper sheet exhibiting excellent adherence to the dryer. A suitable mechanical arrangement for the purpose, including creping on a Yankee dryer is illustrated in U.S. Patent 3,014,832. The sheet as it leaves the press roll tends to expand but, apparently, fibers are retained by the water present so that the fibers remain parallel to a plane.

The sheet on the Yankee was subjected to a surface temperature of about 230°F and the sheet was well dried (3 to 5% moisture) before creping. The sheet adhered well to the dryer and the crepe was of excellent uniformity. The crepe ratio was about 1.7:1.

Under the action of the creping blade the usual hills and valleys developed in the sheet. Additionally, the stiffened fibers released in part and protruded in all directions from the fiber surface. Under the microscope many extending fibers appear as substantially individual fibers while other projecting fibers are in small groups, some of which extend longitudinally of the sheet and bridge a plurality of valleys.

A repeated series of tests based on the papermaking furnish described in detail above, but with varying percentages of the stiff fiber, clearly shows the influence of these fibers upon absorbency in the longitudinal direction. The wicking rate as measured over a 3-inch length of the product increases sharply with the inclusion of 10% of the crosslinked fibers and any substantial proportion, even 5%, is effective. At a 20% inclusion of stiffened fiber the wicking rate in the machine direction doubled; at 40% the wicking rate was quadrupled and at 60% the wicking was 6 times that of a sheet from a furnish of the papermaking fibers having no stiff fiber.

In addition, the wicking rate for the sheet as a whole and the total sheet absorbent capacity increased materially. Also, while the above figures are based on water absorption, oil absorption values for both wicking and total absorption increased to the same extent; for example, the capacity of a sheet to retain oil was doubled by the inclusion of 40% of the stiffened fiber, while the wicking rate for oil was quadrupled.

In the range of 50 to 60% of the stiff fiber the absorbency characteristic is nearly square for most furnishes for both oil and water absorption. While the sheet is more open, aiding wicking, the fiber bridging effect appears to be an important influence in the fluid conduction in the machine direction.

The stiff fibers themselves extend rather heterogeneously and are twisted and curled in various directions though probably a majority of the fibers extend in the longitudinal or machine direction and provide for a considerable wicking action. In many instances the bridging fibers are single fibers. In other instances small groups bundled together, and probably including some papermaking fibers, effect the bridging.

The sheet feel, despite the presence of stiff fibers, is that of a relatively soft nap surface. This web surface does not become as harsh on wetting and redrying as do more tightly bonded sheets. Upon redrying, the sheet retains the protruding stiff fibers.

The product recovers quickly from wet squeezing or pressing. Void volume of the product under light pressures common to the applications in which these tissue products are employed is not only greater than for the usual 100% wood pulp furnish but, importantly, is greater under light pressure, i.e., 0.4 lb/in^2. The introduction, for example, of about 25 to 50% of the stiffened fiber into the plies of a wiping material is effective to lend considerable rigidity to the structure of the sheet and to maintain absorbent capacity even under pressures.

Making Tissue Using Yankee Dryer

D.A. Ely and T.G. McKie; U.S. Patent 3,846,228; November 5, 1974; assigned to Beloit Corporation provide a papermaking machine and process arranged with a view toward eliminating the intermediate steps between the forming wire and Yankee dryer, for making tissue sheet in a more efficient and simplified manner than formerly.

This is accomplished by laying a high speed jet-stream of fibers in an aqueous solution on the up-running run of a Fourdrinier wire, removing a large part of the water from the web while forming and traveling along the forming wire, and then continuing the dewatering and drying by transferring the web directly from the forming wire to the heated drum of a Yankee dryer.

The process is described with reference to Figure 4.1. A looped Fourdrinier or forming wire **10** is shown having a forming run **11** defined by a breast roll **12** at the incoming end of the forming run and a couch roll **13** at the outgoing end of the forming run. The wire **10** is preferably a plastic wire made from an endless plastic mesh which may be of 60 mesh or greater and may be made from polyamide fibers, vinyl fibers or polyester fibers such as nylon, saran, Orlon and Dacron, respectively.

Spaced along the wire on the outgoing side of the breast roll **12** and disposed over a save-all **15** is a conventional stock inlet **16** supplied with pulp from a conventional stock feed means through a pump **17**. The stock inlet **16** may deliver a substantially uniform high speed jet-stream of pulp in the form of tangled fibers in aqueous suspension and lay the pulp on the wire through an inlet slice **19**, disposed over the save-all **15** on the up-running side of the breast roll **12**. The forming wire **10** is adapted to be driven at substantially the relatively high speed jet-stream of the stock, as it is laid onto the wire from the inlet slice **19**.

The save-all **15** positioned just beneath the forming run **11** of the wire and the

FIGURE 4.1: TISSUE-FORMING MACHINE

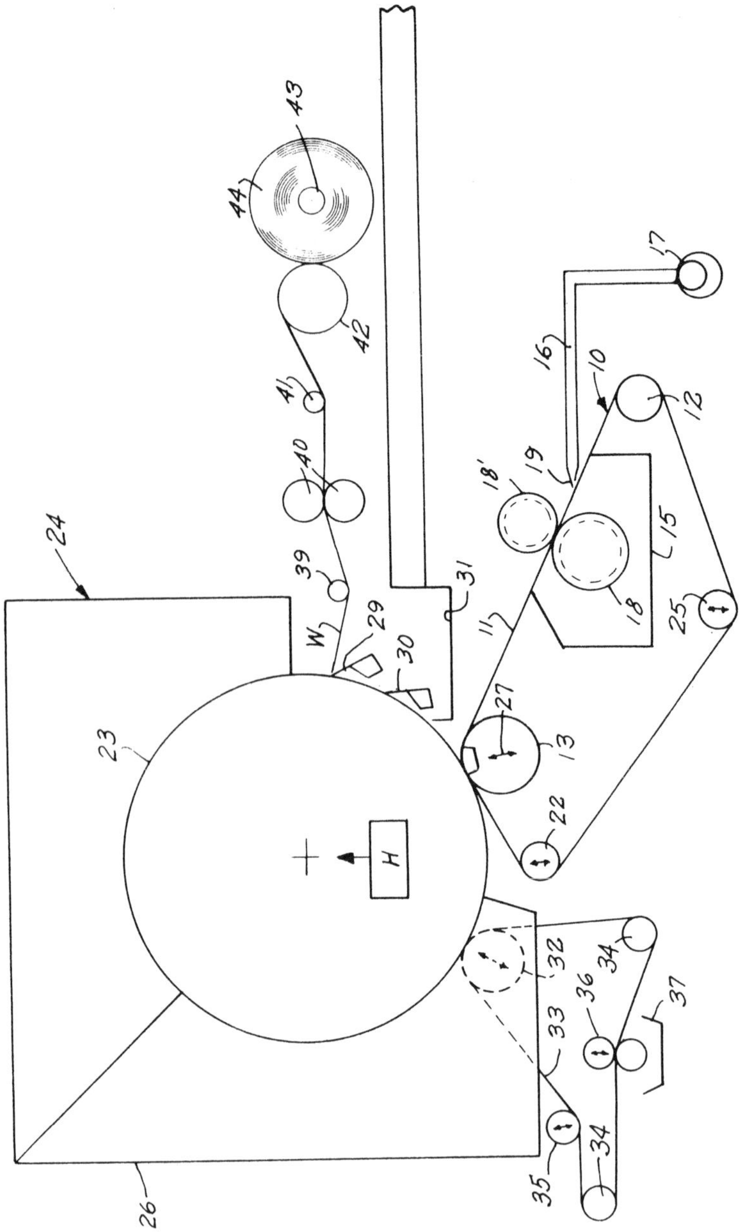

Source: U.S. Patent 3,846,228

inlet slice **19** adjacent the incoming end of the forming run **11** to collect a greater part of the water flowing freely from the stock suspension through the forming run **11** of the wire.

Although the relatively lightweight web forming on the forming run **11** offers nominal resistance to the flow of water therethrough and into the save-all **15**, the weave of the forming run of the wire **11** plus the depositing of fibers in aqueous solution thereon will afford a certain amount of resistance to the water flow through the wire **11** and into the save-all **15**, to the end that there will tend to be a pool of aqueous suspension traveling along the top of the forming run **11** of the wire, which will drain from the web and wire during travel of the wire along the save-all.

Cooperating press rolls **18**, **18'** are shown as being spaced downstream of the inlet slice **19**. The lower press roll **18** is disposed within the save-all **15** and the two press rolls serve to press and dewater the web as it travels along its forming run. The press rolls **18** and **18a** are grooved rolls such as are shown in U.S. Patents 3,198,693; 3,198,694; 3,198,695; 3,198,696 and 3,198,697 to accomodate the free flow of air between the rolls and the wire as the wire passes to the couch roll **13**.

A contact roll **22** trains the wire from the couch roll **13** tangentially of the polished surface of a heated drum **23** of a Yankee dryer **24**. A second contact roll **25** trains the wire back to the breast roll **12**. The contact rolls **22** and **25** may be adjustable to adjust the line of travel of the wire as it leaves the couch roll **13** and to adjust the tension of the wire. The wire is driven at a uniform linear speed approaching or faster than the speed of discharge of the stock from the inlet slice **19**.

The couch roll **13** is shown as a conventional form of suction roll for further dewatering the web to a consistency where it will be pasted to the face of the drum **23** and be carried thereby under a hood **26** extending partially about the drum.

The couch roll may be adjustably moved toward and from the heated drum as indicated by the double-headed arrow **27**, to press a newly formed web or sheet to the face of the heated roll **23** under a preselected pressure.

Where it is desired to imprint a pattern on the web to add bulk to the web the couch roll pressing the web against the surface of the dryer drum **23** may serve to imprint the pattern of the wire **10** on the web.

The hood **26** extends partially about the heated drum to contain the heat of the drum to the sheet and may also serve as a plenum or encase a series of plenum chambers (not shown) directing heated air to the surface of the web as required, and removing moist air from the web to attain a dried product as the sheet or web **W** is creped or doctored from the surface of the heated drum by a doctor **29**.

The doctor **29** may be a conventional form of doctor, doctoring the sheet from the drum **23** and may be followed by a second doctor **30** doctoring broke from the surface of the drum **23** and maintaining the surface of the drum **23** clean. Any broke doctored from the surface of the drum is collected in a broke chute **31**.

Pressure roll **32** is shown pressing a felt **33** against the web on the drum **23**. The felt **33** is trained in an endless path about direction changing idlers **34** and under an adjustable roller **35**. The felt on its return path may also pass between a pair of wringer rolls **36** expressing water from the felt to be collected in a save-all **37**. Pressure roll **32** and wringer roll **36** are conventional.

Where it may be required or desired to imprint a pattern on the web **W** and not be feasible to do this by the plastic wire **10**, the felt **33** may be an imprinting fabric in the form of a mesh and made from a plastic material of fibers such as nylon, saran, Orlon or Dacron.

The doctor **29** is shown as positioned about the drum **23** more than 270° from the point of tangency of the wire **10** to the drum. The dried sheet or web **W** as creped or squeegeed from the drum **23** by the doctor **29** may pass under a contact roll **39**, between a pair of tension rolls **40**, maintaining tension on the sheet as it leaves the drum **23**, under a second contact roll **41** about a roll **42** and coiled or wound on a winder roll **43** into a coil **44**.

Double Diagonally Creped Webs

L.W. Krehnbrink; U.S. Patent 3,476,644; November 4, 1969; assigned to Cincinnati Industries, Inc. describes a method for producing double diagonally creped webs wherein two creping assemblies are provided, each creping assembly comprising a moving creping surface and a creping knife. Means are provided for supporting the creping assemblies one above the other, and means are provided for adjusting the position of the second creping assembly about a vertical axis to locate the second creping surface in a desired path of travel for the web as a control of the stretch characteristics thereof.

In prior art practices, the web has been led to the first creping cylinder in one direction, is caused to travel laterally between the first and second creping cylinders, and then to travel for subsequent treatment steps in a direction offset and parallel to its first path of travel. This is wasteful of factory floor space.

The process provides a creping apparatus in which one creping means is located above the other, with the elimination of adjustable transfer mechanism. This effects a great saving in the cost of the machine and in the factory floor space required.

It also provides a creping apparatus which can be incorporated into a papermaking machine or located to receive the paper off the end of the papermaking machine. In the first instance the creping apparatus of this process may, for example, be located intermediate a series of drying cylinders so as to operate upon the paper web while it still contains some moisture, i.e., that moisture which will be removed by subsequent drying cylinders.

In the second instance the creping apparatus will be operating upon otherwise finished and dried paper, but the location of the creping apparatus to receive the paper as it comes from the paper machine obviates the necessity of forming the paper into rolls, and handling, storing and moving the rolls prior to the creping operation.

Use of Corona Discharge in Dry Creping Method

K.W. Ostermeier; U.S. Patent 3,546,065; December 8, 1970; assigned to Kimberly-Clark Corporation is concerned with the production of creped tissue or creped wadding by the dry-creping process.

In the dry creping of paper the paper web, when quite wet, is pressed onto a drying cylinder and dried to a moisture content usually of about 5 to 10% based on the bone dry weight of the web. The paper adheres more or less firmly to the drying cylinder and is removed from the cylinder by the action of a doctor blade which is urged against the cylinder and serves, in effect, to scrape the paper from the rotating drying cylinder. As an incident of the removal action of the blade, the paper is creped or crinkled.

It was found that by successively applying a corona discharge and a wet web to the creping surface the adhesion or binding of the web to the creping surface, as the web dries, may be significantly increased over that degree of adhesion which the same web would normally display.

The extent of the adhesion is dependent upon the intensity of the corona discharge applied at the creping surface as well as the nature of the web; for a specific web the adhesion is not only increased but is controllably increased. The binding or adhesion may, within limits, be decreased, that is, regulated from a higher degree of bonding to a lower degree though not below that adhesion which would occur between the creping surface and the web in the absence of the corona treatment. The process is described with reference to Figure 4.2.

Referring to Figure 4.2a, a wet tissue web **1**, after being couched in customary manner from a paper machine wire (not shown) and carried to a top felt **2** (illustrated in dash lines) in any suitable known manner, is directed in a substantially continuous manner over guide roll **3** with the top felt into a pressure nip **4**. Nip **4** is formed by press roll **5** and the Yankee drying cylinder **6**.

The web commonly has been partially dewatered prior to passage over roll **3** by various felts (not shown) and by the operation in the nip **4** such that it has a water content of about 60 to 65% by weight immediately after passage of the nip. The wet web adheres to the circumferential creping surface **7** of drying cylinder **6** and is removed from the felt **2** in conventional manner. No release agent need be applied to web **1** as it approaches nip **4** but, if a release agent is present, the effect of the corona still is to increase adhesion of the web.

The sheet or web as it rotates with the dryer adheres thereto to a greater or lesser degree as it dries and loses moisture to the extent that the web as it approaches the doctor blade **8** contains 2 to 10% of moisture by weight. Suitably, the dryer surface has a temperature of between about 85° and 130°C. The blade scrapes the substantially dry web from the creping surface producing the creped or crinkled effect as indicated at **9**. The creped web is then passed over guide roll **10** to a conventional winding mechanism **11** and is reeled at **12**.

The drying and creping procedure thus generally described is known and the application to either the dryer or the web itself of either release or adhesion producing agents is also known.

FIGURE 4.2: USE OF CORONA DISCHARGE IN DRY CREPING METHOD

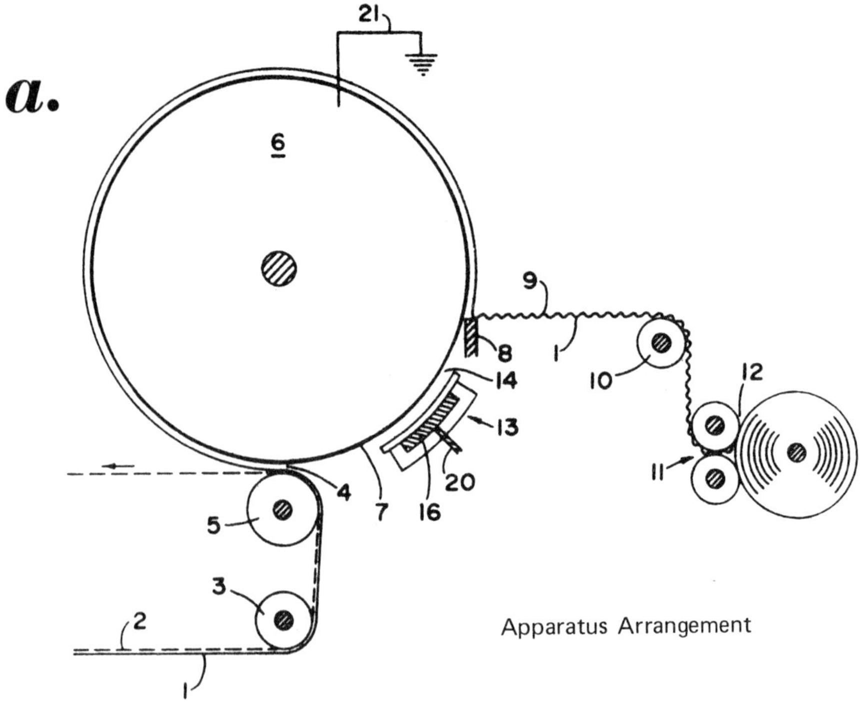

Apparatus Arrangement

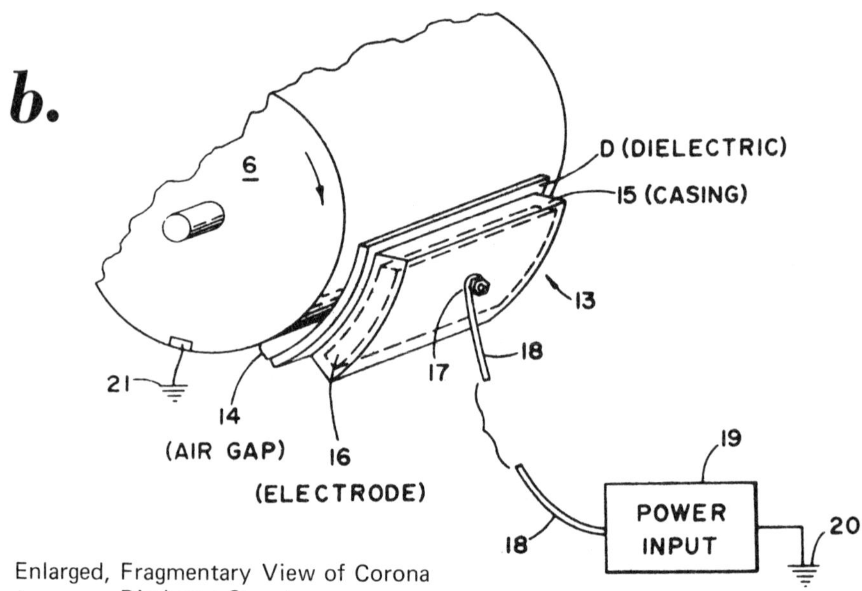

Enlarged, Fragmentary View of Corona Discharge Structure

(continued)

FIGURE 4.2: (continued)

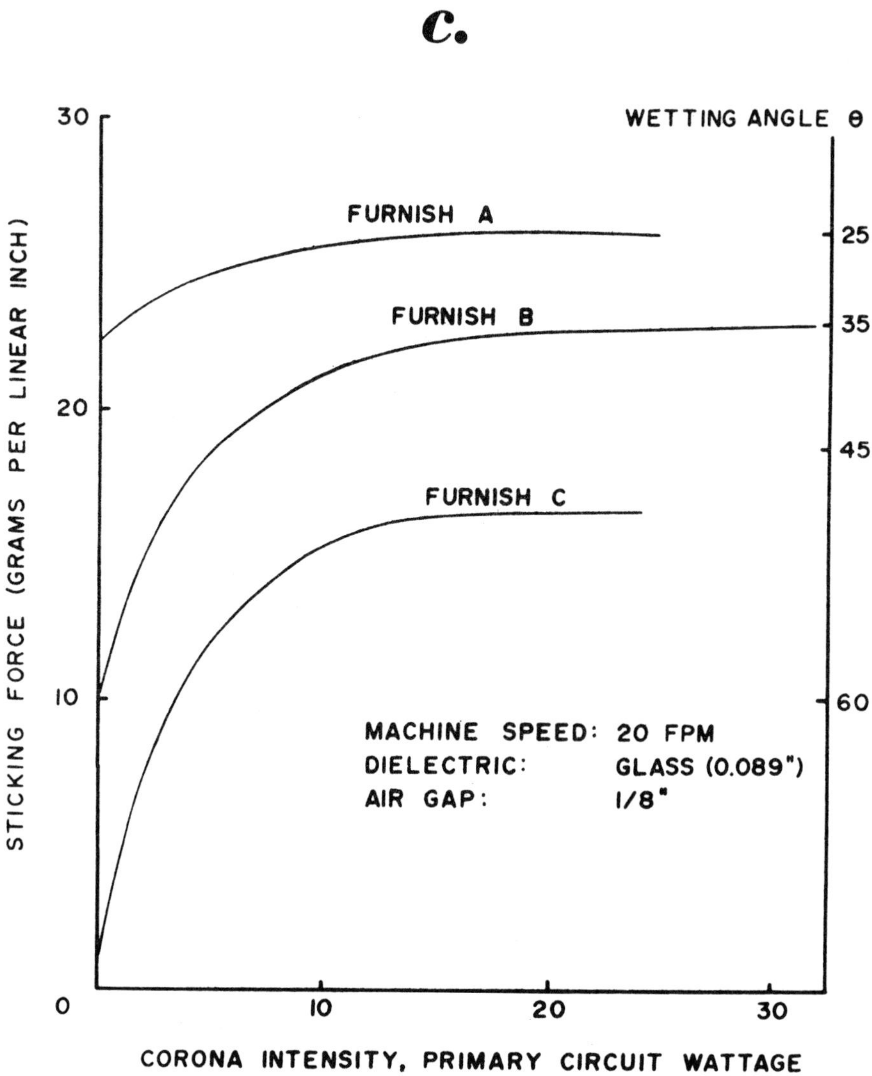

Effect of Corona Treatment on Drier Surface Wetting and on Sheet Adhesion

Source: U.S. Patent 3,546,065

Between blade **8** and nip **4** there is provided an electrode structure generally indicated at **13** for a corona discharge in the air gap illustrated at **14**. The structure **13** extends across the dryer surface (Figure 4.2b) parallel to the dryer axis and is closely positioned to the dryer surface, usually spaced about ⅛" to ¼".

This structure **13** includes a longitudinally extending generally rectangular block **15** of resin serving as a casing and retainer for a metallic electrode **16**. Electrode **16** is suitably of aluminum, steel or brass and is contoured to conform with the dryer surface. Block **15** is suitably of a methyl methacrylate resin. A dielectric material in the form of sheet **D**, preferably in contact with the electrode casing block **15** and spaced from the creping surface **7** of the cylinder, is provided to completely shield the electrode from the cylinder and to inhibit disruptive discharges between the electrode and cylinder. A film of silicone rubber or a sheet of glass serves the purpose as the dielectric material.

A heavy threaded rod engages in the aluminum electrode and extends outwardly of the plastic casing **15** to provide an electrical terminal **17**. Cable **18** connects terminal **17** to any suitable high potential source **19**; such source may include, for example, a high frequency generator supplied at an adjustable input power.

In this instance on equipment illustrating the practice of the process, an input variable between 0 and 10 watts per lineal inch of sheet width at a sheet speed of 20 ft/min has been found satisfactory to produce the required corona at a 1/16" to ⅛" electrode spacing. Potential source **19** is grounded at **20** and the dryer is suitably grounded at **21** through a sliding contact.

Figure 4.2c illustrates findings with respect to the capacity of a corona discharge to increase the ability of a creping cylinder dryer surface to retain webs of various characteristics. The sticking force data for curves **A**, **B** and **C** were obtained by running a sheet of specific character on a slowly moving (20 fpm) dryer surface and withdrawing a determined width of web from the dryer surface just prior to the blade. The withdrawal point was at about the horizontal diameter of the dryer cylinder.

The curve **A** illustrates the effect of the corona discharge at varying intensity on a web of a pulp furnish of 100% long fibered kraft. Webs of 100% kraft commonly adhere well and crepe well. Nevertheless, the adherence is increased significantly from about 22 grams per lineal inch to about 26 grams per lineal inch, or about 18%, by the corona usage.

A furnish having about 60% kraft, 20% sulfite, and 20% groundwood by weight provided the web for the data of curve **B**. As will be readily noted, the sticking force in this instance increased by well over 200%, indicating the greater effect on the more poorly adhering webs. This is further borne out by curve **C** wherein the furnish for the web is about 20% kraft, 60% sulfite and 20% groundwood. The increase in this instance is very considerable percentage-wise. It is thus apparent that the simple use of a corona has a decided beneficial effect on adherence over the whole furnish range. Importantly, such corona usage permits the selection of pulp furnishes and creping techniques which otherwise would not be employed for the purpose.

Figure 4.2c correlates the change in wetting angle of the cylinder dryer surface with the sticking force. It is thus clear that it is the wettability which is altered, and altered without the addition to the surface of chemical agents. Importantly, by simply reducing the intensity of the discharge from the higher values, the sticking force may be reduced from the plateau zones of the curves to lesser values of sticking force. Apparently, this occurs because in operation a new surface is constantly presented for treatment by the corona. In any event control is readily attained so that the creping may be regulated readily.

The process provides that in the creping action the sheet absorbency, softness, bulk and similar properties may be materially improved beyond the conditions obtained when adherence to the creping cylinder is dependent upon drying alone. Additionally, to secure the very high degree of adhesion to the creping cylinder, no creping additives such as adhesives are necessary.

Improving Tissue Paper Softness

E.R. Wand; U.S. Patent 3,592,732; July 13, 1971; assigned to Kimberly-Clark Corporation found that lightweight tissue paper sheets may be adequately worked to provide a soft, limp, smooth sheet of good strength by subjecting the paper sheet to a nip formed between a brush roll and patterned backing roll and imparting a certain speed relationship to the tissue sheet, the backing roll and the brush. The backing roll should be driven at a lower peripheral speed than the sheet and the sheet preferably is at a considerably slower speed than the peripheral speed of the brush roll.

Using a patterned backing roll, the larger the sheet speed relative to the backing roll the greater is the softness attained; additionally, surface smoothness is more readily achieved by moving the brush roll also at high speeds relative to the patterned backing roll, in general two or three to ten times the patterned backing roll speed.

The process is described with reference to Figure 4.3. A sheet or web **1** of wood fiber material is drawn from a supply roll **2** and passed through nip **3** formed by backing roll **4** and brush roll **5** to windup at **6**. Windup rolls **7, 8** rotating in the directions indicated by the arrows support the roll **6** and provide for the winding of the roll in conventional manner.

FIGURE 4.3: IMPROVING TISSUE PAPER SOFTNESS

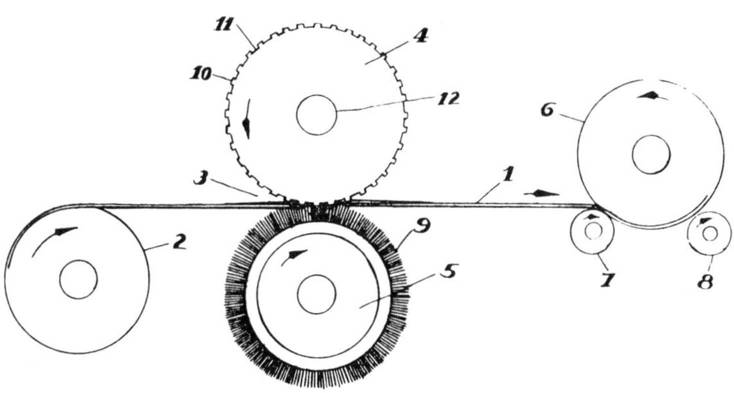

Source: U.S. Patent 3,592,732

The sheet **1** is a single ply of creped cellulosic wadding of about 9 pounds per 2,880 ft^2 dryer basis weight and having a crepe ratio of about 1.20. The sheet was formed by weight of about 50% bleached sulfite pulp, 40% northern kraft and 10% unbleached sulfite. This sheet had an initial tensile strength in the longitudinal direction of about 1,008 grams per 3" width and in the cross direction of about 320 grams per 3" width. The sheet had a bulk density of 1.74 grams per cubic inch, a harshness level of 9.0 at a stiffness of 5.5, and a stretch of 26%.

The supply roll **2**, the backing roll **4** and brush roll **5** are independently driven by conventional means (not shown) in the directions indicated by the arrows. Supply roll **2** is suitably independently driven to prevent stretching of the sheet **1** and, when so driven, the speed of windup at **6** is correlated with the speed of withdrawal. The web speed may be adjusted within a limited speed range between that of the backing roll and the brush roll by varying the tension applied by the unwind roll **2** and the windup roll **6**.

Backing roll **4** is a common type of male embossing roll having pegs **10** uniformly disposed about the roll periphery and also having a shaft **12**. Pegs **10** are of circular configuration and well spaced apart, the pegs having a diameter of about 0.060" and being spaced on centers about 0.110". The roll diameter in this instance is about 10".

Brush roll **5** is suitably about 7½" in diameter and preferably has relatively soft bristles, rows of which are indicated at **9**. The use of soft bristles provides for uniform brush pressure against the traveling sheet **1** and for minimizing of paper strength loss in the softening operation. The brush acts to urge the sheet against the backing roll **4** into depressed zone **11** between pegs or knurlings **10** and to maintain the sheet in contact with the pegs as the sheet slides on the periphery of the backing roll **4** against the sides of the pegs and peg edges.

By pressing of the sheet as the sheet slides on the backing roll, the surface fibers of the sheet are largely released from each other and the body of the sheet to create softness to the touch while simultaneously, the sheet is worked in the depressed zones **11** to break the so-called hydrogen or fiber bonds and provide limpness in the sheet. A very suitable brush roll for the purpose has bristles ¾" long, very closely packed and of about 0.008" diameter.

In practice, employing the specific sheet and arrangement described, the sheet is directed at a speed of about 1,450 fpm to the nip **3**. The peripheral speed of the backing roll is 1,230 fpm and that of the brush roll 3,700 fpm. Under this speed relationship the sheet is readily urged by the brush into the depressed zones **11** and the sheet slides against the lower moving pegs as the sheet leaves the backing roll.

Viewed somewhat differently, the backing roll exerts a drag on the sheet. The combined effect is the production of limpness in the sheet, fiber upsetting resulting in sheet smoothness and softness by the action of the bosses and softening by the action of the brush on the side of the sheet opposite the backing roll.

The sheet product has a final tensile strength in the longitudinal direction of about 808 grams per 3" width, in the cross direction of about 218 grams per 3" width; sheet bulk density is 1.83 gram per cubic inch, and the harshness value 6.0 at a stiffness value of 1.0. Thus, a much more limp sheet of greater softness at small strength loss has been provided. Importantly, no significant embossing of the sheet occurs; if an embossed product is desired, usual embossing procedures may be employed on the softened product.

The brush roll need not be a bristle roll but may be of any type capable of picking on fluffing the fiber surface including wire rolls, toothed rolls and the like.

The speed relationships are of utmost importance, the speed of the backing roll relative to sheet speed and brush roll speed being critical. The backing roll peripheral speed must be less than the sheet speed to provide for sliding of the sheet relative to the backing roll. It is preferred to pass the sheet across the backing roll at a differential speed of about 200 fpm; a decrease in this differential results in a somewhat stiffer product and, conversely, greater speed differentials tend towards more limp materials; a differential of between about 75 fpm and 300 fpm is quite satisfactory for most purposes, and the peripheral speed ranges under such conditions may be between about 425 fpm and 1,425 fpm for the backing roll and 500 fpm and 1,500 fpm for the sheet speed.

Under these generally stated conditions it is preferable to have the brush roll operate at approximately one and a half (1½) to two and a half (2½) times sheet speed, that is, between about 750 fpm and 3,750 fpm, a useful normal brush speed being about 3,700 fpm.

The pressure exerted by the brush roll influences the limpness of the product, the limpness increasing with pressure and with the pattern depth of the backing roll. To obtain surface smoothness, the pressure and pattern depth may be minimal and such conditions are useful when smoothness and softness are desired at low paper strength loss.

TOWELING AND WIPES

Treated Wipes

F.F. Scheuer; U.S. Patent 3,818,533; June 25, 1974; assigned to Alustikin Products, Inc. provides a treated paper, which is disposable, economical and effective in polishing delicate surfaces, such as furniture, cars, metal, plastic and glass, when wiped with any part of this treated paper. The treated paper applies a nondrying film in a thin layer, so as to provide a maximum polish and high gloss without leaving any smearing film.

The paper being treated will absorb sufficient treating material, to render the paper useful for the purpose intended, without altering the structure of the paper if the paper is treated with a nonaqueous liquid vehicle, in which the treating material is substantially uniformly distributed. The treating liquid is self-spreading throughout the entire sheet and is spread evenly over the entire surface of the paper. No separate drying operation is required.

A variety of nonaqueous liquid vehicles are known, which are useful for the purpose of the process. Among these may be mentioned liquid hydrocarbons, glycols and pine oils. The preferred liquid vehicles are the normally liquid hydrocarbons, e.g., normally light liquid hydrocarbon oils, aliphatic naphthas, Stoddard solvents, kerosenes (refined), paraffinic hydrocarbons, natural mineral oil and white mineral oil.

Waxes, scents, pine alcohols, essential oils and/or silicones can be added to the required degree and the resulting admixture remains clear indefinitely without separation. This is true of the treating liquid before it is applied to the nonwoven material or tissue, as well as the treating composition disposed on the paper or tissue.

The relative proportions of the ingredients contained in the impregnating liquid may be varied considerably. Preferably, the ratio of the treating liquid to the weight of treated paper will vary in the range of from about 10 to 70 parts by weight of the treated paper. For example, a multiple of paper sheets, size 12" x 24", requires for end use a 25% treatment. The weight of the paper unit untreated is 10 grams; therefore, 2.5 grams of the solution is applied to any part of this paper unit, which distributes itself throughout the paper evenly in a relatively short time.

The range of wax or combination of waxes added can vary from 1 to 40% of the impregnating liquid, with about 7 to 15% being preferred; the silicone added should not be less than one-tenth of a percent and the range of the odor-producing chemical can vary from about fifteen-hundredth of a percent to about seven-tenths of a percent, all percentages being by weight of the impregnated sheet.

Some waxes which are suitable for use, are carnauba, beeswax, candelilla, paraffin, ceresin, esparto, ouricury and Rezo Wax.

Example: The percentages of the materials indicated are based on the total weight of the treating liquid. 7% of carnauba wax and 3% paraffin wax were mixed with a solution containing 87% hydrocarbon distillate, a colorless, light, nonodorous petroleum distillate, Saybolt viscosity at 100°F 30/35, and 3% silicones, viscosity at 100°F, 100 cs, until fully and uniformly dispersed. Paper treated with this liquid is useful in dusting and polishing.

Multi-Ply Wipe for Use with Organic Solvents

G.H. Meitner; U.S. Patent 3,520,016; July 14, 1970; assigned to Kimberly-Clark Corporation provides a wipe that incorporates a pad of light, low-density, multi-ply creped tissue product to obtain high capacity to absorb and hold organic solvents or water, and a gauze cover material to improve the structural integrity of the pad particularly at its edges, to screen pad fibers from escape to the surface being cleaned, and to enhance the appearance of the wipe.

The wipe comprises a multi-ply pad and a gauze fabric envelope (preferably cotton threads without binder) wrapped around the pad and having sewn side edges and unfastened ends which project beyond the edges of the pad. It has been found that a wipe so constructed and utilizing a multi-ply creped tissue product for the pad which has a low density within a preferred range provides a high capacity to absorb organic solvents or water.

These are used for cleaning surfaces such as the metal skin of aircraft preliminary to painting, and will readily release such liquids for their cleaning action upon the application of hand wiping pressure. The structure of the pad is described with reference to Figure 4.4.

FIGURE 4.4: CONSTRUCTION OF INNER PAD OF WIPE

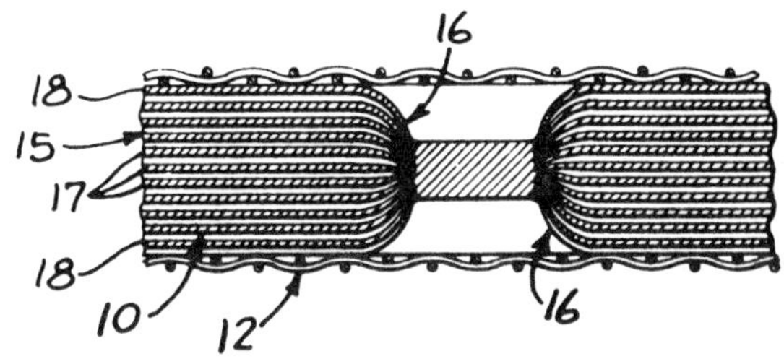

Source: U.S. Patent 3,520,016

The product for the pad **10** is comprised of a multiplicity of layers **15** of creped cellulosic tissue united by uniformly distributed embossments **16**. To provide a pad **10** having superior solvent absorbency and release characteristics, the layers of creped cellulosic tissue preferably have a low density within a range of between about 0.02 g/cm^3 and 0.06 g/cm^3.

Such a product for the pad has been made by employing nine plies of coarsely creped tissue **17** (crepe ratio about 1.8) having a dry, uncreped basis weight of between 4.5 and 6 pounds per ream of 3,000 square feet and assembling the nine plies of coarsely creped tissue **17** with cover sheets **18** of finely creped tissue (crepe ratio of about 2.70) having a dry, uncreped basis weight of about 4.5 pounds per ream of 3,000 square feet, so as to produce a composite product with a density of about 0.026 g/cm^3 (1.62 lb/ft^3).

Multiple layers of coarsely creped tissue sheets having a basis weight within the range of from about 4.5 to 6 pounds per 3,000 square feet and a crepe ratio within the range of 1.6 to 2.2 pounds also are suitable for the pad product without also employing such finely creped tissue cover sheets **18**, although there is a small sacrifice in appearance when such a product is used for the pad because of the less uniform surface texture of the coarsely creped tissue **17**.

The plies of coarsely creped tissue **17** and finely creped cover sheets **18** are united by embossments **16** which cover 2 to 10% of the area of the pad and, together with the sewn edges, insure the structural integrity of the composite assembly of pad and gauze envelope.

The wipes have superior solvent absorbency and releasibility characteristics. It has been found that rectangular 8" x 9" wipes approximately ¼" thick hold up to 68% more methyl ethyl ketone solvent and up to 154% more water than all gauze wipes of the same weight.

In addition to having higher solvent capacity per wipe, the wipes of the process also release solvent better under hand wiping pressure than all gauze pads and hence are effective to carry more solvent to the surface being cleaned. They are also more effective for carrying dirt and other foreign matter away from the surface being cleaned.

To provide a wipe which after the solvent has dried will leave the cleaned surface substantially free of residue extracted by the solvent from the wipe, the elements of the wipe are made from materials characterized by substantial absence of constituents extractable in organic solvents including methyl ethyl ketone, acetone, carbon tetrachloride and toluene, or water. To accomplish this end, the multiply product for the pad **10** has been made from kraft or sulfite pulps, or mixtures thereof, with or without wet strength resins added, which meet the requirements of low extractables in the specified solvents (less than 0.5%).

Paper Having Plurality of Crests and Depressions

R.W. Schutte, K.W. Britt and A.L. McConnell; U.S. Patent Reissue 27,453; Aug. 1, 1972; assigned to Scott Paper Co. describe a disposable industrial wiper having high affinity for oil and grease, which wiper is strong in both machine direction and cross-machine direction and which has sufficient caliper or bulk to provide a generally thick sheet so that the person using a single sheet of the product is satisfied with the feel thereof and is not tempted to use a plurality of sheets for a single cleaning operation when a single sheet would suffice.

The absorbent paper or wiper preferably is formed from base-stock which has high wet-strength and is highly absorbent. A good product for use as an industrial wiping tissue is obtained from long-fiber base-stock weighing approximately 16 pounds per ream and which is strong and porous and which also preferably contains a wet-strength agent.

The base-stock is passed through a press or forming machine which includes a lower roll and an upper roll each rotatably mounted on its own axis and supported in juxtaposition to the other roll. The rolls preferably are interconnected through suitable gears so that they rotate in synchronism.

It is preferred that the upper roll be supported whereby the surface thereof is in spaced relation to the surface of the lower roll (the gap being a few thousandths of an inch, more or less) so that the rolls do not come in contact, whereby to reduce wear on the roll-surface and prevent the entire surface of the base-stock from being compressed.

The rolls are designed generally to compress a predetermined amount of the surface area of the base-stock and to stretch portions of the base-stock between the compressed areas beyond the elastic limit of the base-stock whereby more or less to separate the fibers from one another and thereby increase the porosity of the base-stock. The compressed area is less than 50% of the total paper area and preferably is approximately 10%. The rolls are not smooth but each is formed with a generally nubby surface comprised of many projections and depressions.

The projections may be formed by machining or engraving the roll surfaces, or may be provided by wrapping a screen of wire mesh around a smooth cylindrical roll or in any other suitable manner. Rolls formed by wrapping wire screen around a smooth cylindrical roll have been found to be satisfactory. A useful operative wire screen has a 12 x 12 mesh and is made of wire 0.047 inch in diameter. For instance, the mesh may vary from a mesh of 6 x 6 to 24 x 24 with wire diameter ranging from 0.023" to 0.072". Each roll thus formed has a plurality of spaced high spots and intervening depressions, the high spots on one roll being generally in alignment with the depressions on the other roll. The 12 x 12 mesh wire pattern provides 72 projections, or high spots per square inch of roll surface and a like number of depressions in the same area.

The amplitude of the undulations in each roll is greater than the thickness of the base stock and when the base stock is fed between the rolls, it is pushed by the high spots on one roll into the depressions on the other roll.

In one modification the clearance between the crest of each high spot and the bottom of the juxtaposed depression exceeds the thickness of the base stock and the base stock is not compressed in these areas. However, along the slopes of the undulations of the so-deformed paper the base stock is stretched beyond its modulus of elasticity and is deformed.

The stretching referred to above may, in some cases, exceed the tensile strength of the paper and tearing or shredding of the paper will occur and apertures will appear in the paper. The apertures preferably are defined by fibrous strands and not by sharp shear lines. These fibrous strands attract fluid into the paper-body and increase the absorbency of the paper as well as the rate of absorbency of the paper.

It is to be understood that shredded fibers adjacent the apertures are not indispensable because a multitude of so-formed narrow slits or apertures will attract and hold fluid therein by capillary action, and this is the case even if the apertures are defined by straight, clean edges (as when the base stock is sheared by a sharp instrument).

It is desired that the clearance between the surface of the rolls in the areas to be shredded be less than the thickness of the base stock so that in these areas the juxtaposed rolls grind or crush the fibers and shred the paper as by a tearing action.

The paper of the process is characterized by having a high compressibility factor. That is to say, the ratio of the caliper or overall thickness of the product (measured from the crests on one side to the crests on the other side) in an uncompressed state, to the product thickness after pressure is applied, is substantially high.

For instance the paper may be approximately 0.028 inch thick under substantially no pressure and after compression at 2 pounds per square inch it is reduced in thickness to approximately 0.006 inch. Thus the compressibility ratio is approximately 4½ to 1. This is substantially greater than the compressibility of the base stock alone which, at a pressure of 2 pounds per square inch, is reduced from approximately 0.009" to approximately 0.0055".

Relatively Inextensible Center Ply Bonded to Highly Extensible Outer Plies

The process developed by *C.B. Kemp; U.S. Patent 3,953,638; April 27, 1976; assigned to The Procter & Gamble Company* is used to produce a multilayered absorbent wiping product exhibiting superior bulk and strength, both wet and dry, by imprinting a regulated pattern of reinforcing resin on both sides of a web of tissue paper which is relatively inextensible, and bonding a web of highly extensible, creped tissue paper to each side of the central ply by means of the reinforcing resin contained on the central ply at mating embossment points on a pair of synchronously driven, hard-surfaced embossing rolls.

The wiping product is comprised of a relatively inextensible central ply of cellulosic tissue paper having outer plies of highly extensible creped cellulosic tissue paper intermittently adhered thereto. The central ply is formed from a wet strength paper having less than about 10% stretch. An intermittent reinforcing pattern of a suitable polymer resin is imprinted on each surface of the central ply.

The resin reinforcing pattern covers between about 15 and about 55% of the total surface area on each side of the central ply. The inextensibility of the central ply causes the unadhered areas of the outer plies joined to the central ply by means of the resin to expand in the Z-direction when the laminate becomes wetted, thereby providing superior wet bulk caliper and impression.

The resin reinforcing pattern imprinted on each side of the central ply serves several purposes. First, the wet and dry tensile strength of the absorbent wiping product is increased due to the reinforcing effect on the central ply. Second, the machine direction to cross-machine direction tensile ratio of the finished product may be adjusted in any desired direction by varying the composition of the resin, the imprint pattern, and the amount of resin applied.

Third, the resin reinforcing pattern imprinted on both sides of the central ply permits bonding of the highly creped, outer tissue plies to the central ply solely by means of pressure exerted between mating embossment points on a pair of synchronously driven, hard-surfaced embossing rolls.

Example: A laminate structure was produced utilizing a cellulosic tissue center ply having a basis weight of approximately 10 pounds per 3,000 sq. ft. at approximately 8% crepe prior to imprinting. Conventional processing and papermaking techniques were employed in producing the center ply. Approximately 20 pounds per ton of Kymene S-2064, a wet strength resin, and 10 pounds per ton of carboxymethylcellulose were added to the furnish which was comprised of 60% softwood kraft and 40% hardwood sulfite. The center ply was produced generally in accordance with the process of U.S. Patent 3,301,746 on a 36 x 30 square mesh fabric.

Both outermost plies were produced with a basis weight of approximately 16 pounds per 3,000 sq. ft. at approximately 30% crepe. Conventional processing and papermaking techniques were employed in producing both outermost plies. Approximately 20 pounds per ton of Kymene S-2064 wet strength resin and 10 pounds per ton of carboxymethylcellulose were added to the furnish which was comprised of 70% softwood kraft and 30% hardwood sulfite. Both outermost plies were produced generally in accordance with the process of U.S. Patent 3,301,746 on a 31 x 27 semitwill fabric.

The center ply was imprinted on each side with Elvace EP-1962, a selfcrosslinking acetate/ethylene copolymer resin at a rate of approximately 2.1 pounds of solids per 3,000 sq. ft. per side. A diamond-shaped imprinting pattern comprised of lines having a thickness of 0.020 inches and arranged 15 lines per inch in the machine direction by 7 lines per inch in the cross-machine direction was employed. No attempt was made to print the diamond pattern on one side of the central ply in register with the diamond pattern on the opposite side of the central ply. The reinforced central ply material was stored in roll form for a period of approximately 4 weeks at ambient temperatures before the final converting operation was performed.

The outermost plies were laminated to the central ply by means of the reinforcing resin utilizing mating knob-style embossment patterns having 64 knobs per sq. in. on each side of the structure. The area embossed by the knobs constituted approximately 17% of the structure's total surface area on each side. The laminate structure was calendered by passing it through the nip between a pair of calender rolls adjusted to a minimum gap of about 10 mils to meet the target caliper of approximately 30 mils.

Individual sheets of the product were then cured by inserting them into an oven in which 350°F hot air was being circulated for a period of 2 minutes. Basis weight of the finished laminate structure was approximately 39.9 pounds per 3,000 sq. ft. The product exhibited these characteristics:

Dry Caliper at 80 g/in^2 loading	31.6 mils
Wet Caliper at 80 g/in^2 loading	26.4 mils
Ratio of Wet Caliper:Dry Caliper	83.5 %
Dry Tensile Strength	
Machine Direction	1,724 g/in
Cross-Machine Direction	1,387 g/in
Wet Tensile Strength	
Machine Direction	796 g/in
Cross-Machine Direction	653 g/in
Dry Stretch	
Machine Direction	8.0 %
Cross-Machine Direction	8.0 %
Wet Stretch	
Machine Direction	4.0 %
Cross-Machine Direction	6.0 %
Dry Tensile Ratio	
Machine Direction:Cross-Machine Direction	1.242
Wet Tensile Ratio	
Machine Direction:Cross-Machine Direction	1.218
Tearing Resistance	
Machine Direction	78 g/ply
Peel Strength	1.73 g/in^2
Handle-O-Meter	
Machine Direction	59 g
Cross-Machine Direction	50 g
Absorptive Capacity	15.0 g H_2O/g product
Rate of Absorption*	36.2 sec

*Time in seconds required to absorb 0.10 ml of distilled water.

Adhesive Applied to Preselected Portions of Web

R.E. Small, J.J. Bradley, C.J. Sanders, H.J. Slawny and S.R. Brown; U.S. Patent

3,673,052; June 27, 1972; assigned to International Paper Company and Paper Converting Machine Company, Inc. describe paper toweling utilizing two sheets.

According to the process, at least one of the sheets is caused to travel with and partially wrap on a plate cylinder, with the adhesive being provided on the surface of the sheet remote from the cylinder. The cylinder advantageously is equipped with an outstanding pattern of an open perimetric configuration, i.e., diamonds, squares, etc. The adhesive-equipped sheet is thereafter integrated or united in the adhesive-equipped areas with a second sheet or web.

It is advantageous to tension one of the webs, preferably the unprinted web (with adhesive) during the time it is integrated with the other web or webs. This then results in a product wherein the previously tensioned web puckers or contracts when relieved of tensioning stress so as to develop an extremely desirable quality in the resultant product both from the standpoint of appearance and structure. The product has an essentially quilted appearance which not only is attractive but also provides pockets for the take up of the moisture, important where, for example, the product is used as toweling.

With particular reference to Figure 4.5a, the numeral **10** designates generally a fibrous web which is seen to pass between embossing rolls **11** and **12**. The embossing rolls may be engraved steel rolls with patterns of embossments provided therein. For example, the pattern may be a repeat type having from 10 to 200 projections or protuberances per square inch, each raised to a height of from about 0.01 to about 0.05 inch whereby the sheet **10**, when subjected to such action results in having an embossed surface of from 10 to 75% of its total area.

The web **10** then passes over an idler roll **13** and about a plate cylinder generally designated **14**. The plate cylinder **14**, as can be seen in Figure 4.5b, includes a rubber covering **14a** which overlies the steel interior roll **14b**. The web **10** is seen to be in partial wrapping engagement for travel with the roll **14** and, in the course of so traveling, picks up a layer of adhesive **15**. This occurs on the portions of the web which are in contact with the raised portions **16** provided on the rubber covering **14a** of the plate cylinder **14**.

The adhesive is derived from a glue bath **17** and is picked up therefrom by means of a rubber covered roller **18**. The roller **18** (which constitutes a fountain roll) transfers the adhesive from the bath **17** to a transfer roll **19** from which it is applied to selected portions of web **10**.

The web **10**, equipped with adhesive in preselected portions (based upon the configuration of the rubber printing plate portions **16**), is integrated with a second web **20**. The web is seen to have passed between embossing rolls **21** and **22** which are optional though advantageous depending upon end usage. Whether surfaced for embossment or not, the rolls **21** and **22** are run at a slower speed than succeeding rolls engaging the web **20** so that the web **20** is tensioned.

The webs **10** and **20** are integrated in the nip defined by the cylinders **14** and **23**. The united web after traveling a portion of the arcuate length of the roll **23** are caused to pass between a nip **24** defined by rolls **23** and **25**. Thereafter the multi-ply web **26** is directed to other structure (not shown) for reeling, perforation, etc. In some instances the roll **25** is structured with a plate identical to that of **14a** and registered to contact the webs only where glue has been applied by **14a**. This insures a good glue bond without crushing of the embossing where glue has not been applied.

A number of distinct advantages arise from the manner of glue application just described. For one thing, the glue does not come in contact with the plate cylinder **14** so there is no problem of cleaning. Further, in previous arrangements, glue had been applied to the so-called plate cylinder and tended to attack the sticky-back which was employed to adhere the plate **14a** to the steel roll **14b**. For example, sticky-back is used here at the inner face designated **14c** in Figure 4.5b (between the plate **14a** and the cylinder **14b**).

FIGURE 4.5: ADHESIVE APPLIED TO PRESELECTED PORTIONS OF WEB

a.

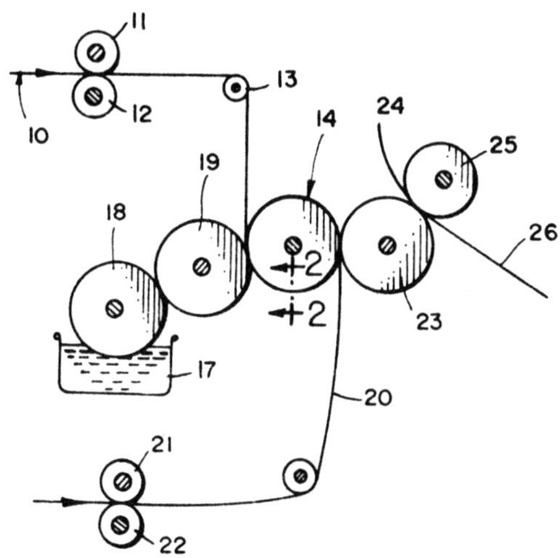

b.

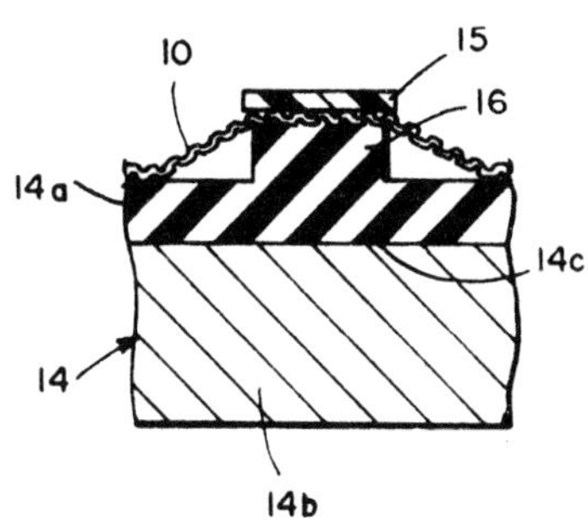

(continued)

FIGURE 4.5 (continued)

c.

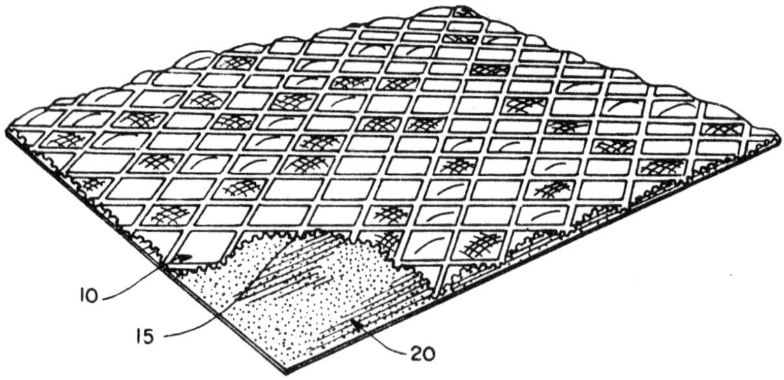

(a) Schematic Elevational View of Apparatus
(b) Enlarged Sectional View Along **2–2** of (a)
(c) Perspective View, Partially in Section of Product.

Source: U.S. Patent 3,673,052

The product can be seen in Figure 4.5c wherein there is a distinct puffiness or quilted appearance in the resulted multi-ply sheet **26**. The upper ply or web **10** is caused to puff or quilt by virtue of the contracted nature of the under ply **20**. The web **20** has fewer fibers per unit area in contact with the adhesive as at **15** than the web **10**, this resulting from the fact that the web **20** was stretched at the time the bond (via the adhesive) was made between the sheets **10** and **20**.

Crepe Papers with Interposed Plastic Film

Y. Yoshioka; U.S. Patent 3,925,127; December 9, 1975; assigned to Gomei Kaisha Touyo Shoji, Japan describes four bulky papers, useful as towels which have a plastic film laminated therein to provide sufficiently high bulkiness and strength as well as good dimensional stability.

The first bulky paper is prepared by inserting a plastic film between a pair of crepe papers, and passing the assembly through the nip between a pair of heated embossing rolls to cause a partial adhesion of the plastic film with the crepe papers. The first bulky paper is characterized by its high bulkiness which increases with the crepe ratio of the crepe papers, and also by a good dimensional stability owing to the fact that the plastic film is interposed between and bonded to the crepe papers.

The second bulky paper is produced by supplying moisture to the crepe papers of the first bulky paper to extend the crepe, whereupon it is dried at a temperature which does not modify the plastic film. The second bulky paper is characterized by a higher bulkiness as compared with the first bulky paper as well as its dimensional stability which is as good as that of the first bulky paper.

The third bulky paper is produced by thermally treating the first bulky paper to cause a shrinkage and/or melting of the plastic film which is interposed between the crepe papers. The third bulky paper is characterized by its bulkiness which is higher than that of the first bulky paper as a result of the shrinkage of the plastic film. The melting of the plastic film causes it to fuse partially with the crepe papers, thus losing the film configuration, and removing the rattling which inherently occurs due to the presence of the plastic film.

This avoids objectionable noises offensive to the ear when it is used as a towel or towelette. As compared with the first bulky paper, the dimensional stability is as good.

The fourth bulky paper is produced by supplying moisture to the crepe papers of the first bulky paper to extend the crepe, followed by a thermal treatment to cause a shrinkage and/or melting of the plastic film interposed between the crepe papers. The fourth bulky paper is characterized by a bulkiness which is higher than any of the first to third bulky papers. In addition, the plastic film is melted and thus converted into a network configuration, whereby the rattling is eliminated as in the third bulky paper with concomitant advantages. Compared with the first bulky paper, the dimensional stability is as good.

One of the most important features of the crepe papers used is its crepe ratio, which must be in excess of 10%, preferably in excess of 40%.

The plastic film which may be used includes films of thermoplastic synthetic resins, such as, polyolefins, polyvinyl chloride, polyvinylidene chloride, polyamide and polyester. Polyethylene film is preferred in view of its low cost and low melting temperature.

Generally, the plastic film has a thickness between 3 and 100 microns. The thinner the thickness of the film, the more pliable the bulky paper becomes, but the contribution of the plastic film to increasing the strength of the bulky paper is reduced.

Example 1: A pair of crepe papers formed of 100% pulp and having a crepe ratio of 50% and a basis weight of 25 g/m^2 is used, and a 15 micron thick polyethylene film manufactured by the T-shaped die process in the same size as the crepe papers are laminated therewith. One of the embossing rolls comprises a steel roll having a diameter of 15 cm and a length of 50 cm and formed with a relief pattern of cruciform projections measuring 5 mm long and 2 mm wide, with a spacing of 1 cm on centers, and the other embossing roll is formed by a hard rubber roll having a diameter of 15 cm and a length of 50 cm.

The assembly comprising the crepe papers with the interposed polyethylene film is passed between these embossing rolls under a pressure of 10 kg/cm^2 while maintaining the rolls at a temperature of 170°C and rotating them with a rotational speed of 21 rpm, thereby causing an adhesion of the polyethylene film to the crepe papers at spaced points to produce the bulky paper of the process. The bonding takes place in the cruciform configuration at a rate of nearly 700 per 30 cm^2.

Example 2: The bulky paper obtained in Example 1 is sprayed on both surfaces with water at a rate of 100 ml/m^2 and dried for about four minutes by a hot air drying machine maintained at a temperature of about 130°C, also producing a bulky paper.

Example 3: The bulky paper obtained in Example 1 is subjected to a thermal treatment by the hot air drying machine maintained at about 130°C for three minutes, thereby also producing a bulky paper.

Example 4: The bulky paper obtained in Example 1 is sprayed with water on its both surfaces as in Example 2, and subsequently subjected to a thermal treatment similar to Example 3, again producing a bulky paper.

HYGIENIC PAPERS

Tissue with Controlled Wet Breakup for Recirculating Toilets

The tissue developed by *K.M. Enloe and J.T. Gresham; U.S. Patent 3,798,120; March 19, 1974; assigned to Kimberly-Clark Corporation* comprises a coherent self-sustaining web made from conventional absorbent fibers of papermaking length, but which have selected areas of the web additionally bonded by a wet strength binder. The areas treated by the wet strength binder encompass a major percentage of the total web area and are preferably in the form of multiple patches arranged in a closely spaced patchwork pattern in which the patches are isolated from each other by narrow, binder-free web portions, of a minimal width.

In the fiber-aligned or machine direction of the web these binder-free portions are preferably only slightly wider than the average fiber lengths. In the transverse direction the binder-free portions may be even narrower, again because of normal fiber alignment. However, in modern papermaking equipment, fiber alignment can be varied to increase cross direction strength. In such event, the transverse direction spaces may be slightly wider than fiber length too.

When tissue sheets of this nature are deposited in flowing or circulating water, as in flushing the toilet, the action of the water will quickly weaken the narrow unbonded portions, causing the sheet to break up into individual patches of a size corresponding to the size of the original wet-strengthened areas. When such tissue is used with a recirculating toilet, the released patches do not clog the filter media or its operating parts, as a large mass of individual fibers will do, since the patches have sufficient body to be moved around by the recirculating aqueous medium along with the rest of the solid wastes. The toilet may thus remain in serviceable condition for a longer time period than when ordinarily easily disintegratable tissue, or overall wet-strengthened tissue is used.

The wet strength binder may be selected from urea-formaldehyde, melamine-formaldehyde, polyamide-epichlorohydrins, vinylamide polymers, polyethylene imines, and other similar wet strength resin materials.

The resin is applied in an amount sufficient to prevent the treated areas of the web from disintegrating into component fibers for at least 24 hours and preferably for a period of several days. Usually an amount of between about 0.2 and 0.8% solids on a fiber weight basis, is applied.

The wet strength binder may be applied in the form of a clear solution, or it may contain selected dyes, pigments, or other colorations to suit various needs, or for aesthetic reasons.

The overall size of each of the wet strengthened areas has been found to be most satisfactory when the area of each patch is in the neighborhood of about 900 square millimeters, with the binder free spaces between wet-strengthened portions having a width of about 3 millimeters in longer fibered webs, and being of lesser width in shorter fibered webs. However, the treated tissues will still minimize clogging of the filters, if the size of the individual wet strengthened areas is as large as about 10,000 mm^2 or as small as about 25 mm^2.

Preferably the wet-strengthened areas comprise from about 75 to about 95% of the total web area with the higher percentages being preferred.

The general style of the pattern used in applying the wet strength binder to the web may be of an almost infinite variety. It may assume the form of repeated geometric patterns including polygons such as triangles, rectangles, hexagons, octagons and the like, or it may be in the form of various regular or irregular ornamental designs such as trademarks, service marks, or other distinctive insignia as may be elected.

The tissue may be in the form of a roll divided into perforated segments, folded sheets, interfolded sheets, of single or multiple plies.

Example: A creped tissue web of absorbent facial grade stock having an uncreped basis weight of about 7.6 pounds per 2,880 sq. ft. was printed on a flexographic press with an 8% aqueous solution of a wet strength resin, then dried and cured. The fibers used were about 80% softwood and 20% hardwood. The resin used was a polyamide-epichlorohydrin. The pattern used was large spaced squares where the spaces between printed areas were staggered. The sides of the squares were about 25 mm long. The resin-free portions were about 3 mm wide. The amount of resin solids pickup in the treated tissue was about 0.5% by weight.

About 560 4" x 4" sheets of the dried and cured web were deposited in the aqueous medium of a recirculating toilet, which contained approximately 8 gallons of fluid. Upon agitation during the recirculating process, the sheets rapidly broke up into squares of approximately the same size as the printed pattern. Very few individualized fibers were noted. No excessive hangup of fibers or web pieces were found in the filtration system, even after extended operation.

Other sheets were dropped into a beaker of water and agitated to simulate the action of a recirculating toilet for a period of about one minute and then allowed to stand. These sheets were broken up into small squares by the agitation and these squares retained their integrity after agitation. The broken up material was left in the beaker, agitated and checked again at 12 hours, at 24 hours, and then after several days. In each instance, the integrity of the pieces was substantially maintained.

Two plies of the same sheets (as above) were assembled in perforated roll form. Similar tests were run on these sheets with substantially the same results.

Yeast Impregnated Tissue for Use with Septic Tanks

The process developed by *C. Willey, W.R. Willey and C.R. Willey; U.S. Patent 3,645,769; February 29, 1972* has a particular advantage for use in those areas where septic tanks or cesspools are required in waste disposal systems. The successful operation of a septic tank or cesspool is dependent on the continuous supply and nourishment of bacteria therein that results in a chemical reaction to cause disintegration of the waste products deposited in the septic tank.

It has been found in many instances that toilet tissue deposited in septic tanks does not react immediately to the chemical reaction brought about by the bacteria located in the septic tank and thus tends to clog the septic tank, resulting in faulty operation thereof.

It is commonly accepted that live yeast in particular metabolic products thereof supply growth factors for the cultivation of bacteria and the process incorporates this principle and combines yeast with the conventional toilet tissue to promote the disintegration thereof.

The tissue is prepared in a conventional manner. The yeast is impregnated into the finished toilet tissue, the yeast being in a dry powder form and blown under air pressure into the finished paper. It is seen that when the toilet tissue with the dry viable yeast impregnated therein is introduced into the liquid such as found in a septic tank, the addition of the liquid to the yeast will begin promoting bacterial growth to thereafter cause disintegration of the tissue in which it is impregnated.

It has been found that only a very small quantity of the yeast material is required for impregnation into the toilet tissue to achieve the desired results. The amount of the yeast impregnated in the tissue will be in the range of 0.5 to 2.5% by weight with respect to the tissue. A specific quantity of the dry yeast as impregnated into the finished paper tissue that has been found to yield the desired results is 1.7% by weight with respect to the toilet tissue. It is understood that the above indicated percentages of the yeast materials as introduced into the tissue may vary in accordance with the weight and the thickness of the tissue employed.

A modified form of the process involves the impregnation of the yeast into the finished paper, while the yeast is in a wet or liquefied condition. This may be accomplished by forming a paste of the yeast and water and applying it to the tissue. The yeast is then allowed to dry on the paper at low temperature before the paper is wound in roll form.

Special care must be taken so that the yeast is not killed in the course of the process. It is well known that yeast would be killed by heating the impregnated paper to a temperature of about 56°C, thus care must be taken that in drying or any other steps in the process after the yeast has been added that the temperature is not raised to a point where the yeast would be killed.

Avoiding Mechanical Compression Until Sheet Is 80% Dry

The product described by *D.L. Shaw; U.S. Patent 3,821,068; June 28, 1974; assigned to Scott Paper Company* is a soft, absorbent, creped, fibrous web formed by deposition from an aqueous slurry of principally lignocellulosic fibers. The web is formed and dried to a predetermined dryness level without being subjected to mechanical compression. At dryness levels beyond this predetermined dryness level, further interbonding of the fibers by papermaking bonds to a significant extent will not occur upon mechanical compression of the web. The predetermined dryness level is chosen to produce a very soft, low density web.

Further reduction of papermaking bonds, and thus softening of the web is accomplished by creping the web from a dryer surface. The web is adhered to the dryer surface through the use of a creping adhesive, chosen to adhere the web to the dryer surface without adding significant interfiber bonding to the web. The fibers in the web are bonded together almost exclusively by papermaking bonds formed upon drying of the web. The web thus formed has exceptional softness and surprisingly adequate strength when compared to other known products relying upon papermaking bonds for their strength.

The predetermined dryness level is preferably at least 80% fiber consistency, that is, containing no more than 20% moisture by weight, but the web may be rewet to as low as 70% fiber consistency upon application of the creping adhesive to the web. The web is subjected to a modest pressure when it is applied to the creping surface, increasing web density and formation of papermaking bonds to a slight extent when the web has been rewet to 70% dryness, and much less so when the web has been rewet to no wetter than 80% fiber consistency.

The adhesive can be applied to the web directly or to the drum first, by either spraying or printing with a rotogravure roll. The adhesive is selected for its ability to adhere the web to the drum and not for adding strength to the web. In practice, the adhesive is a material which because of its nature or the amount used adds no significant strength to the web, less than 15% of the total web strength, for example. Examples of suitable adhesives are Accostrength 95 and Accostrength 86, Creptrol 190 and animal glues.

The most common adhesives suitable for this purpose are in aqueous solutions, and therefore, the web is rewet somewhat upon application of the adhesive. In the most preferred method, the web is more than 90% dry upon application of the adhesive, and it is rewet to a degree which reduces the web dryness to no less than 80%.

Example 1: The web was formed from an aqueous slurry of 80% kraft softwood pulp and 20% kraft hardwood pulp, which was passed through an open gap and formed into a sheet. The sheet was through-dried without application of mechanical compression to 98% fiber consistency by passing heated air through the web. The sheet was then adhered to a Yankee dryer drum with a transfer roll exerting about 150 psi pressure to the web and with the aid of an adhesive applied to the web by gravure printing in a repeating hexagonal pattern, 40 mils in length, 7.5 mils wide, and 35 microns deep, from a copper roll with a 0.002 inch thick chrome outer layer.

The adhesive was an aqueous solution of 3% Accostrength 95, 1% National Starch latex 5329, and 0.34% Rycofax 618. The dryness of the web after printing was 76%, and the web had picked up 0.7% adhesive solids. The sheet was then creped and lightly calendered for bulk control. The creped sheet was converted into tissue rolls. The converted paper had the general properties shown on the following page.

Basis weight	16.8 lb/2,880 ft^2
Bulk	0.197 in/24 sheets
Tensile (MD)	16.1 oz/in
Stretch (MD)	16.6 %
TEA (MD)	0.955 kg-m/m^2
Tensile (CD)	4.6 oz/in
Stretch	8.4 %
TEA (CD)	0.370 kg-m/m^2
L (MD)	4.0 cm
L (CD)	4.65 cm

Example 2: A web was formed from an aqueous slurry of 80% kraft softwood fibers and 20% kraft hardwood fibers refined to a freeness of 560 ml. To the aqueous fibrous slurry, 0.2% Quaker 2000 (based on bond dry fiber weight) was added to reduce the interfiber bonding capacity of the fibers. The fiber suspension was formed into a sheet on a Fourdrinier wire and through-dried without application of mechanical compression with hot air to 95% fiber consistency.

The sheet was then adhered to a Yankee dryer drum by spraying an aqueous solution of 1% Accostrength 86. The amount of chemical pickup to the web from the application of the creping adhesive was less than 1%. The amount of pressure exerted on the web by the transfer roll was about 150 psi. The sheet was then creped and converted into bathroom tissue rolls, and had the following general properties.

Basis weight	17.4 lb/2,880 ft^2
Bulk	0.188 in/24 sheets
Tensile (MD)	23.6 oz/in
Stretch (MD)	19.3 %
TEA (MD)	2.20 kg-m/m^2
Tensile (CD)	5.9 oz/in
Stretch (CD)	6.8 %
TEA (CD)	0.364 kg-m/m^2
L (MD)	4.4 cm
L (CD)	4.9 cm

The TEA-to-stiffness ratio is obtained by first measuring the TEA (tensile energy absorption) of a given specimen of sheet material in accordance with the TAPPI Test, T494 SU–64, in both the machine direction (MD) and the cross-machine direction (CD), in kilogram-meters per square meter, with the exception that a jaw spacing of 2 inches rather than the 8 inches recommended by TAPPI is used because of the particular nature of the product, some of which have lines of perforation which must be avoided.

This test method is not a TAPPI standard, but is suggested by TAPPI as the most suitable method to date. The stiffness of the product is then measured by subjecting the specimen to the test set forth in TAPPI standard test, T451 M–60, in both the machine direction and the cross-machine direction, to determine its effective overhanging length (critical length) denoted as L in centimeters. The stiffness of the product is proportional to the cube of the effective overhanging length.

Temporarily Waterproof Paper Product

The process of *J.I. Lindgren, P.E.C. Udden and B.A. Wennerblom; U.S. Patent 3,922,434; November 25, 1975* relates to a temporarily waterproof paper article comprising a fibrous substrate capable of disintegration in water and a covering particulate layer produced from a powder of a water-soluble polymer.

Particularly suitable polymers are water-soluble cellulose derivatives such as methylcellulose, hydroxyethylcellulose, ethylhydroxyethylcellulose and carboxymethylcellulose, (sodium salt).

Preferably the polymer in a 2% aqueous solution should have a viscosity of at least 50 cp (centipoises) at 20°C, measured with a Brookfield viscosimeter at 50 rpm.

There are several ways of combining the carrier material and the water-soluble polymer in particle form. One consists in suspending the polymer particles in a liquid, in which the polymer material is insoluble, after which the suspension is applied to the surface of the carrier material and the liquid is caused to evaporate. This way of application is termed coating.

A second means of combining a carrier material and polymer in particle form consists in the direct application of dry polymer to the carrier. This way of application is called dusting. The dusting may be carried out as a simple application of finely distributed polymer to the surface of the carrier. In case the carrier is of a porous nature a certain amount of polymer powder will then adhere purely mechanically. Any surplus of powder may, for instance, be brushed away.

In connection with an application to the surface of the kind mentioned, it may be of advantage that the carrier be a little moist, in which case the binding to the polymer layer is improved.

A deeper distribution of the polymer particles in the carrier may, for example, be obtained by sucking, pressing or vibrating the powder into the carrier. The dusting methods have certain drawbacks. It is difficult to obtain a uniform covering layer and, dusting problems readily occur. This can be avoided by using an electrostatic covering method. In that case, for instance, the carrier is moved continuously over an earthed roller, while negatively loaded polymer powder is being blown against the paper.

A third possibility of combining the carrier material with a polymer in particle form consists in adding the polymer in connection with the manufacture of the carrier or in partially transforming the carrier material into a water-soluble polymer by chemical transformation. In this connection this is termed simultaneous manufacture.

One way of adding the polymer in connection with the manufacture of the carrier consists in, for example, mixing cellulose powder and pulverulent water-soluble polymer, adding a small amount of water and then compression molding the mixture.

Due to the unique combination of properties of a temporary waterproofness with a certain amount of permeability to gas and, in case a suitable carrier is selected, also with the capability of breaking up in water, products obtained in connection with the process includes diapers, sanitary napkins, insertions for bed-pans, toilet paper and wash cloths.

Example: A web of crepe paper, so-called tissue paper, having a basis weight of about 20 g/m^2, was moved through a vat filled with finely ground NaCMC of the quality SCA (Svenska Cellulosa Aktiebolaget) Cellufix 3000 spec. Then about 7.5 g of powder, per m^2 and side, of the paper, adhered to the paper. In order that there should not be too much dust from the paper in connection with the continued handling thereof the least adhering CMC-powder was removed by slight vibrations. There was left about 4 g of CMC-powder, per m^2 and side, of the paper.

For reasons, readily understood, a paper covered in this way is not particularly waterproof as far as one layer is concerned. If, however, three such layers are placed on each other, it will be found that this combination of layers does not allow water to penetrate therethrough for many hours.

INSULATION PAPERS

IMPROVING DIELECTRIC PROPERTIES OF PAPER BY REPLACEMENT OF MONOVALENT CATIONS WITH DIVALENT CATIONS

Paper has been and remains a very suitable material for dielectrics in capacitors and for cable insulation applications, but modern capacitor and cable designs require ever higher quality paper dielectrics and insulation with lower dielectric loss (dissipation factor, tan δ) over a wide range of voltage stress and temperature. Significant improvements have been made by both the pulp and papermakers in recent years and continued efforts are being made to obtain papers of lower loss characteristics.

The mechanism of dielectric loss or conduction in a heterogeneous substance such as paper or liquid-impregnated paper is extremely complex and is not fully understood. Dielectric loss is generally attributed to rotation of dipoles present, and to oscillation of free ions. Ionic material is always present in paper to some extent.

The monovalent cations are very mobile and possess relatively low affinity for organic functional groups and, because of these characteristics, contribute substantially to dielectric losses in paper. Trivalent and tetravalent cations, although least mobile, do not combine readily with these functional groups, because of steric hindrance. The divalent cations appear to have the best combination of affinity for carboxyl groups and diminished mobility. Accordingly, replacement of monovalent cations with divalent cations is beneficial in reducing dielectric loss in papers.

The process of *W.B. Cranford, A.J. Petricola and M.J. McLeod; U.S. Patent 3,475,269; October 28, 1969; assigned to Canadian International Paper Company, Canada* relates to the production of papers having improved dielectric properties. More particularly, it relates to a process leading to condenser and cable grade papers having a diminished susceptibility to dielectric loss, i.e., to dissipation of energy within the dielectric due to a sinusoidal electric stress, which is characterized by expression ωCV^2 tan δ, where ω is 2π x frequency, C is the capacitance,

V is the voltage, and tan δ is the power factor.

It has been found that large reductions in the dissipation factor (tan δ) of pulps can be achieved if the pulps intended for capacitor and cable insulation papers are treated in aqueous solutions with compounds of barium, calcium, magnesium, and strontium. The acetates, hydroxides, sulfates, and other salts of these cations show promise.

$Mg(OH)_2$, MgO, $MgSO_4$ and $Ba(OH)_2$ have been found very effective. The hydroxide, in addition, neutralizes the deleterious presence of the worst offenders in terms of dielectric loss, i.e., the hydrogen ions. (Since one of the main conduction mechanisms is believed to be that of protonic conduction, a pH over 7.0 is desirable in eliminating hydrogen ions.) This type of treatment has been found to be effective even against the dielectric loss which would be expected in conditions of prolonged pulp aging at high temperature (about 120°C).

Example: Pulp (raw stock) is washed with water and then subjected to an acid soak to remove most of the metallic contaminants (ash). The addition of SO_2 at pH 3.0, 40°C, and 2½% OD for 30 minutes accomplished this, though other acids can be used. The pulp is then well washed to remove the acid.

Next magnesium hydroxide is metered into the pulp slurry at a pH of about 6.0 to 9.0, preferably above 7.0 and best about 8.0. If an acid salt such as $MgSO_4$ is used, the initial pH will generally fall below 7.0. The pulp can have a wide range of consistency, but preferably from about 2 to 6% OD and can have a temperature in the range of room temperature to boiling. 2½% OD and 40°C have been found effective.

For the best results, provision should be made to obtain uniform blending of the pulp and the magnesium hydroxide and to allow enough time, e.g., up to 120 minutes, for the reactions involved to be completed. These ion exchange reactions involve the functional groups (carboxyl groups of uronic acids and phenolic OH^- groups) associated with γ- and β-celluloses and lignin and can be represented as follows:

$$2R\cdot COOH + MgX_2 \rightleftharpoons (R\cdot COO)_2Mg + 2HX$$

where R is a polysaccharide matrix and X is an acetate, chloride, hydroxide, sulfate, etc., anion. Finally, the pulp should be thoroughly washed to remove the excess, unreacted magnesium compounds, etc., and is ready for conventional handling.

It has been found that when two portions of pulp, one dealt with in accordance with the process, as described above, and the other merely acid-soaked and washed as described above, are compared in terms of dissipation factor, tan δ, the pulp subjected to the process has such factor reduced by 40% or more.

BORON-FILLED PAPER FOR INCREASED THERMAL CONDUCTIVITY OF DIELECTRIC PAPER

Due to its high dielectric strength, paper is often employed as an electrical insulator in the manufacture of various types of equipment. However, since paper is a poor thermal conductor, its use is generally limited to those applications where heat dissipation is not a problem.

J.W. Tereshko; U.S. Patent 3,562,101; February 9, 1971; assigned to Union Carbide Corporation describes a paper which is characterized by both a high dielectric strength and a high degree of thermal conductivity.

This is accomplished by incorporating boron nitride into paper such that the paper's thermal conductivity is increased while its excellent dielectric properties are maintained. The process comprises preparing an aqueous slurry of cellulosic pulp from water and fibers which are formed by fibrillating wood, rags, and plant fibers, then dispersing boron nitride particles substantially uniformly throughout the cellulosic pulp slurry and depositing the pulp and the particles which have been homogeneously intermixed on a substrate to form a paper mat and processing the mat through a conventional papermaking process to produce boron-nitride-filled paper.

The quantity of boron nitride which is present in the paper may vary from 10 or less to 80% by weight of the paper. The greater the quantity of boron nitride, the greater will be the thermal conductivity of the paper. However, practical considerations, such as the adverse effect on the strength of the paper, cause it to be preferable to produce a product having between 25 and 65% boron nitride by weight of paper. These preferable ratios may be obtained by employing a quantity of boron nitride particles weighing about 50 to 300% by weight of the pulp slurry to form the boron nitride slurry.

For most effective and efficient results when combining the dielectric paper and the boron nitride, the latter should be in a powder form having a particle size of from 0.01 to 44 microns. Larger particles are difficult to disperse during processing and the boron nitride is not generally dispersed well in the final product.

As commercially produced, boron nitride is often adversely affected by moisture. To further enhance the properites of the boron-nitride-filled paper, the boron nitride particles may be coated with a resin material before being joined with the aqueous pulp slurry. Suitable resins are thermoplastic resins such as polyvinyl chlorides and acetates. A typical resin of this type is Bakelite or a vinyl solution Resin VYHH (Union Carbide). These resins are neutral, nonoxidizing, non-heat-reactive, have a low water vapor permeability and low water absorption capability. This latter characteristic is of course most desirable for any resin which is employed.

Thermosetting resins such as epoxies and phenolics may also be used to coat the boron nitride particles. Suitable epoxies are epichlorohydrin-bisphenol A and its copolymer. Phenolics are more difficult to apply to the particles and so phenolic varnishes are preferable resins to employ.

The resin is normally dissolved in a suitable solvent to obtain a thin solution, and boron nitride powder is mixed in the solution to obtain a thick slurry. The mixing may be carried out in a paddle blade or bread mixer. The mix is then heated to evaporate the solvent and/or cure the resin. The resultant mass is pulverized to obtain the desired particle size for addition to the aqueous pulp slurry.

As an example of this procedure, the aforementioned VYHH resin was dissolved in equal amounts of methyl ethyl ketone and toluene. The boron nitride particles were then added to the solution. Additional solvent of the same proportions was added as needed to maintain the heavy slurry consistency. After all of the boron nitride was homogeneously mixed, the solvent was allowed to evaporate until cracking occurred. Then the mixture was cured at 120°C for two hours using a 30°C/hr rate of temperature rise. After curing, the resin-coated material was reduced to a powder. The resin amounted to about 2% by weight of the powder.

In order to test the effectiveness of this type of coating, boron nitride having the resin coating applied was placed in an atmosphere having a 90% humidity. Analysis after a testing period of 8 hours showed that the coated boron nitride had a water content of 1 mg/g of boron nitride, whereas the water pickup of uncoated boron nitride was six times greater.

In addition to the aforementioned moisture-resistant coating procedure, the boron-nitride-filled paper of the process may be coated with boron nitride. The boron nitride coating serves to supplement any small voids which occur due to the paper not being completely filled with boron nitride during processing. Furthermore, the coating insures a complete paper-to-paper contact when the boron nitride paper is overlapped during assembly.

Example: (a) The Production of Paper Containing Uncoated Boron Nitride – A pulp fiber furnish was prepared by refining 100% unbleached softwood kraft to a Canadian Standard Freeness of 300. 50 lb of this pulp furnish was diluted with water to give a stock consistency (percent solids) of approximately 1.5. In a separate container, a slurry of filler material was prepared by adding boron nitride powder to water which contained methanol as a wetting agent. The particle size of boron nitride used was in the 0.2 micron to 1.0 micron range. In order to assure good wetting and a uniform dispersion of the boron nitride powder in the water, a high-speed stirrer was employed to vigorously agitate the water while the boron nitride was being added thereto. Tap water was used as the diluent throughout the experiment with no pH adjustment.

The so-treated pulp fiber furnish and filler material were then mixed together and continuously stirred for about 30 minutes. Water was added to the resultant homogeneous mixture to adjust its consistency to the degree required for processing it on the papermaking machine. Boron-nitride containing paper was then produced by processing the foregoing boron nitride pulp fiber furnished on a laboratory-size Fourdrinier papermaking machine.

(b) Properties of Paper – A number of tests were carried out on a series of boron-nitride-filled paper which was produced by employing the process described above. The following table illustrates the properties of the papers which were produced.

As indicated in the following table, the thermal conductivity of the boron nitride-filled paper increased as much as 100% over the thermal conductivity of the paper having no boron nitride therein. It is to be further noted that the dielectric properties of the paper were also adversely affected although these data are not presented in the table.

Percent wood pulp	Percent BN	Percent BN in paper	Tensile force, lbs.	Weight of 8 x 8" sheet, g. approximate	Average thickness per sheet, cm.	Thermal conductivity, cal./cm. sec./°C.
100	0	0	15.88	2.5	.104	1.20×10^{-4}
80	20	5.2	11.95	2.5	.108	1.28×10^{-4}
75	25	9.1	9.37	2.5	.108	1.31×10^{-4}
50	50	24.7	4.75	2.5	.094	1.43×10^{-4}
40	60	39.2	1.66	2.5	.122	1.60×10^{-4}
33	67	56.8	1.65	7.5	.212	2.64×10^{-4}
25	75	62.5	3.94	5.0	.227	2.33×10^{-4}

COATED ELECTRICAL INSULATING PAPER

Paper intended for use in electrical insulating applications where high dielectric strength is required, as in capacitors for example, is made on Fourdrinier paper machines, using specially purified, very highly beaten kraft woodpulp. Such paper is essentially film-like in character, i.e., it is substantially free of holes and of internal voids, and has extremely low permeability to fluids.

However, unlike films made from homogeneous systems, e.g., cellophane and cast or extruded plastic films, the surface of capacitor paper is rough or toothy on a microscopic scale. This fine scale roughness is an important advantage in the use of paper as electrical insulation, since it provides passages for the ingress of liquid impregnants between the layers of insulation. However, whenever two microscopic valleys on opposite sides of the paper happen to occur at the same position, a thin spot results which represents a point of dielectric weakness and potential dielectric failure.

K. Suter and J.H. Matthews; U.S. Patent 3,903,352; September 2, 1975; assigned to Kimberly-Clark Corporation are concerned with increasing the dielectric strength of electrical insulating paper without however increasing its thickness. Maintaining the paper thin is important should it be used as a dielectric spacer in a capacitor, since as is known the smaller the dielectric spacing between the electrodes of a capacitor, the greater will be its capacitance.

It was found that the application to the paper surface of dilute solutions of certain water-soluble, film-forming polymers produces dramatic increases in the dielectric strength of the paper.

The improved paper of a given total basis weight, including both fiber and surface coating, is superior in dielectric strength to conventional paper of the same weight, composed entirely of fiber.

The material must be insoluble in the liquid dielectrics commonly used to impregnate paper insulation, such as chlorinated biphenyls, mineral oil, etc; it must not itself cause an intolerable increase in the dielectric loss factor of the paper; and it must be substantially free of alkali metals, because the presence of such metals increases the power factor of the paper.

Since it was found that all otherwise appropriate materials contain excessive quantities of sodium, a practical process was developed for removing alkali metals from certain potentially useful, water-soluble, film-forming materials.

The film-forming material must be one which is insoluble in water within one temperature range but which dissolves within some different temperature range. For ease of processing, both of these temperature ranges preferably lie below 100°C but this is not essential. By insoluble is meant that the material must not become highly swollen (gelatinized) or tacky, but it must be permeable to water so that diffusion of ions from the solid into the water phase can take place.

Materials having this critical characteristic permit the easy removal of alkali metals by simply washing them with water of low alkali metal content, maintaining the temperature within the range in which dissolution cannot occur. The efficiency of the washing process is improved by adding very small amounts of harmless cations, such as hydrogen ions or alkaline earth ions, to the wash water.

These cations do not impair the electrical properties of the film-forming material as do alkali metal ions, and they serve to displace alkali metal ions from cation-exchange sites in the material, thereby greatly reducing the amount of washing required to effect the desired reduction in alkali metal content. It is advantageous to provide the cations in the early stages of the washing process by adding magnesium sulfate or sulfuric acid to the wash water in the amount of about 0.01 to 0.1% by weight of the water.

The washing process is also facilitated by having the material to be washed in a finely divided state (high surface-to-volume ratio) so that the rate of diffusion of ions from the solid into the liquid phase will be rapid.

The washing can be carried out by any established processes for washing finely divided solids, such as repeated batchwise dispersing, settling, and decantation; continuous washing and centrifugal decantation; and countercurrent washing with rotary filters. Whatever the technique, extraction must be continued until the alkali metal content of the film-forming material, which will typically be about 1,000 ppm before washing, has been reduced to no more that 200 ppm, and preferably, to no more than about 20 ppm.

Four materials which meet the criteria for use as film formers are methylcellulose, starch, polyvinyl alcohol, and protein. Methyl esters of cellulose are manufactured as Methocel. Preferred are grades having viscosities between 10 and 25 cp (2% aqueous solution, 20°C) and a degree of substitution between 1.64 and 1.92. This type of methylcellulose is insoluble in organic solvents and has the unusual property of being insoluble in hot water but readily soluble in cold water.

Accordingly, the washing process with this material is carried out at a temperature of about 80°C, and the slurry of washed methylcellulose is then chilled to 0° to 10°C in order to obtain a gel-free solution. After the methylcellulose has dissolved, solubility is retained if the solution temperature is increased, up to about 35°C, and the viscosity of the solution is reduced. The optimal temperature for application to the paper is about 30°C.

The starches most useful are the oxidized and the chemically substituted types, such as hydroxyethylated starch, acetylated starch, etc., which are commonly used as surface sizes and coating binders in the manufacture of printing and writing papers.

Preferred are the low and medium-viscosity grades having viscosities below about 200 cp (10% aqueous solution, 20°C). These starches are insoluble in cold or warm water but dissolve when cooked at about 90°C for 20 to 30 minutes. Once dissolved, the starch remains in solution on cooling, and the solution temperature can be lowered for application to the paper. The optimal temperature for application can be anywhere between 30° and 90°C, depending upon the solution concentration and the viscosity of the starch.

Polyvinyl alcohol is available in types which are soluble in water over the whole temperature range between 0° and 100°C, but the type employed is the so-called fully hydrolyzed which has about 98% or more of the original acetyl groups removed by hydrolysis. This type of polyvinyl alcohol is similar to starch in its solubility characteristics, i.e., it is insoluble in water at moderate temperatures but dissolves on cooking at 80°C, depending upon the molecular weight and the number of residual acetyl groups. Preferred are the products with viscosities between about 25 and 125 cp (4% solution, 20°C).

The preferred protein is purified soy protein, commonly known as alpha-protein. The low-viscosity version of this material is the most suitable. Unlike the other three materials described above, protein requires the addition of ammonium hydroxide to a pH of about 8 to 9, as well as heating to about 50°C or higher in order to obtain solubility.

All four of these water-soluble, film-forming materials are about equal in their ability to increase the electric strength of insulating papers. However, polyvinyl alcohol and protein increase the dissipation factor of the treated paper a small amount and, for this reason, these two materials are less suitable then methylcellulose and starch in applications such as the making of capacitors, where this property of electrical insulation is critical.

Methylcellulose, in addition to increasing the electric strength of paper, has a unique ability to suppress corona discharge and to prolong the life of paper dielectrics under conditions of elevated temperature and high electrical stress.

The water-soluble, film-forming material can be applied to paper by any of the commonly used techniques of coating or surface-treating paper, such as size press coating, blade coating, roll coating, air knife coating, etc. It may be applied either on or off the paper machine; the former is preferred for reasons of economy.

The coating material may be applied to only one side of the paper, or to both sides. Generally, treatment of both sides is preferred in order to maximize the effect of the treatment and to avoid curl problems.

The amount of surface treatment which is required to provide significant improvement of electric strength is at least about 2%, on a dry weight basis, of the total weight of the treated paper, and the preferred range is from about 3 to about 10%.

Example 1: Methylcellulose having a degree of substitution of 1.64 to 1.92 was washed by the process previously described until its sodium content was reduced to 12 ppm. It was then dissolved in water to form a 4% by weight solution by cooling to 5°C. This solution had a viscosity of about 120 cp at 20°C. The solution was applied to a web of capacitor tissue by means of a conventional size press located on the paper machine. The basis weight of the resulting treated paper product was adjusted to obtain the same weight obtained prior to the application of the methylcellulose solution. Both the treated and untreated papers were supercalendered to a thickness of 0.5 mil and a density of 1.0 g/cc.

It was found that the regular capacitor tissue which had a basis weight of 5.3 pounds per 2,000 square feet and a thickness of 0.51 mil had a dielectric strength of 1,540 v/mil when tested according to ASTM D-202-70. The treated paper, having a dry basis weight of 5.3 lb/2,000 ft^2 of which 3.8% was methylcellulose, and a thickness of 0.50 mil, had a dielectric strength of 1,850 v/mil.

Example 2: Treated paper was produced as described in Example 1 except that a 3.8% by weight solution of methylcellulose was used to coat a paper which, after supercalendering, had a thickness of 0.4 mil. The paper produced prior to coating had a basis weight of 4.2 lb/2,000 ft^2, a thickness of 0.41 mil and a dielectric strength of 1,770 v/mil. The coated paper, also having a basis weight of 4.2 lb/2,000 ft^2 of which 3.7% was methylcellulose and a thickness of 0.40 mil, had a dielectric strength of 2,080 v/mil.

The coated and uncoated papers were also used to construct conventional askarel-impregnated 5.0 μF capacitors which were stressed to failure with DC voltage. Each capacitor was made by winding metal foil electrodes interleaved with paper, placing the wound body in an enclosure, and evacuating the enclosure and filling it with askarel. The capacitors made with the regular, untreated paper showed an average breakdown voltage of 1,690 volts DC while the units made from the treated paper showed an average breakdown voltage of 2,140 volts DC.

Example 3: Treated paper was produced as described in Example 1 except that a hydroxyethylated starch was prepared by washing to a sodium content of 10 ppm by the method previously described and then dissolved in water by heating to 90°C for 30 minutes to produce an 8% by weight solution.

The temperature of the solution was adjusted to 50°C before application to the paper, at which temperature the viscosity was 65 cp. The paper produced prior to application of the starch solution was found to have a basis weight of 4.0 lb per 2,000 ft^2 and a thickness of 0.4 mil and a dielectric strength of 1,520 volts per mil. The treated paper had a dry basis weight of 4.1 lb/2,000 ft^2, of which 7.8% was starch, a thickness of 0.4 mil, and a dielectric strength of 2,030 v/mil.

Conventional 5.0 μF capacitors impregnated with askarel as described above, were produced from both treated and untreated papers and electrically stressed to failure. The units made using untreated paper had an average breakdown value of 1,600 volts DC while the treated units showed an average breakdown value of 1,910 volts DC.

FRICTIONALLY CALENDERED POLYOLEFIN PAPER

H. Miyamoto, I. Igami, M. Uehara, H. Fujita, H. Itoh and Z. Kan; U.S. Patent 3,957,573; May 18, 1976; assigned to Dainichi-Nippon Cables, Ltd. and Mitsubishi Rayon Co., Japan provide a process for producing a polypropylene paper suitable for use as an insulating paper, which has improved air impermeability, oil-resistance and mechanical strength. This comprises forming a sheet out of polyolefin fibers containing at least 70% by weight of polypropylene fibers having a denier of at least 0.5 and a birefringence of at least 2.5×10^{-2} and frictionally calendering the sheet at a frictional ratio of at least 15% and at a temperature in the range from 90° to 160°C. There are variations to the process.

While in all methods the fiber sheet comprises at least 70% polypropylene having a denier of at least 0.5 and a birefringence of at least 2.5×10^{-2}, in one modification a sheet composed of a fiber mixture of such polypropylene fibers with fibers of a different polypropylene, polyethylene or a mixture of such a different polypropylene and polyethylene is frictionally calendered.

In a second modification polypropylene fibers having a denier of at least 0.5 and a birefringence of at least 2.5×10^{-2} are sheeted in the form of a bicomponent fiber with polyethylene, e.g., such polypropylene is simultaneously extruded with separately melted polyethylene to give a two-component fiber formed of a polypropylene fiber joined to a polyethylene fiber, or a fiber formed of such polypropylene is coated with polyethylene. In a third modification, a sheet composed of a mixture of polypropylene fibers having a denier of at least 0.5 and a birefringence of at least 2.5×10^{-2} with bicomponent fibers as described are subjected to frictional calendering.

In any variation, a polypropylene insulating paper with useful air-impermeability, oil-resistance and mechanical strength is obtained by fractional calendering at a temperature of from 90° to 160°C and at a frictional ratio of at least 15%.

However, where significant proportions of polyethylene are present, superior results are obtained at less severe calendering conditions within the above range, i.e., close to the 90°C-15% lower boundary, while to obtain optimum results with fiber systems containing less than about 5% polyethylene, one preferably frictionally calenders at a temperature of at least 115°C and at a frictional ratio of at least 20%.

The polypropylene fibers are firstly formed into sheets, using conventional machinery for the fabrication of dry nonwoven fabrics such as a garnet machine or a random webber, or by a wet papermaking method using a binder, or by a dry papermaking method using a random sheeter.

However, a wet papermaking process which generally comprises cutting the polypropylene fibers into a cut length of from 0.5 to 20 mm, preferably from 3 to 15 mm, and dispersing such short-cut fibers in a dispersing medium such as water or an organic solvent is particularly preferred since the process provides a sheet having a uniform air impermeability. The term short-cut fibers used herein refers to the fibers having a cut length of from 0.5 to 20 mm, preferably from 3 to 15 mm.

When water is used as a dispersing medium in the wet papermaking process, short-cut polypropylene fibers are dispersed together with a binder in water, and the resulting dispersion is subjected to a papermaking process to obtain a sheet where the fibers are temporarily adhered together by the binder.

The sheet is then heated at a temperature greater than the shrinkage temperature of the polypropylene fiber (generally about 130°C) but lower than the melting point of the polypropylene fiber to give a sheet having a high wet strength due to the entangling of the fibers during shrinkage. The sheet is subsequently washed with water to remove binder remaining in the sheet. As the binder used the so-called wet-heat melting types are preferred, in an amount of from about 10 to about 30 parts by weight of binder per 100 parts by weight of the polypropylene fiber.

When an organic solvent is used as a dispersing medium in the wet papermaking process, short-cut polypropylene fibers are dispersed in an organic solvent in which the polypropylene is not dissolved at room temperature but in which the polypropylene is soluble at high temperature, for example, tetralin, decalin, xylene, methylene chloride, or mixtures thereof, and the resulting dispersion is subjected to the papermaking process.

The wet web thus obtained is then heated to partially dissolve the polypropylene fiber in the solvent remaining in the web, to melt the fibers sufficiently to adhere them to each other to provide excellent mechanical strength to the sheet.

When the polypropylene fiber is heated at a temperature greater than the shrinkage temperature of the polypropylene fiber (usually about 130°C) but less than the melting point of the polypropylene, the temperature at which the polypropylene dissolves in the above organic solvent rises to some extent.

A sheet having excellent mechanical strength can be obtained from a mixture of such heat-treated polypropylene fibers and untreated polypropylene fibers by subjecting the mixture to a wet papermaking process and then heating the resulting wet web at a temperature at which the previously heat-treated fiber does not dissolve but the untreated fiber dissolves sufficiently to adhere to the treated fiber.

The sheet obtained in the above manner is dried, and then subjected to a frictional calendering. The polypropylene fiber sheet is passed between a pair of rolls, each of which rotates at a peripheral speed different from the other, i.e., friction calender rolls. Such a frictional calendering is carried out at a frictional ratio of 15%, or more, and at a sheet temperature in the range of from 90° to 160°C, more preferably, at a frictional ratio of 20%, or more, and a sheet temperature in the range of from 115° to 160°C, whereby the sheet undergoes a shearing force due to the difference of the peripheral speed of the rolls.

Generally, a frictional ratio of less than about 300% is used with good results, more preferably less than about 100%. The polypropylene fibers in the sheet, in particular the fibers located at the surface of the sheet, are deformed and partially fill the interspace among the fibers in the sheet, thereby increasing the air impermeability of the sheet. In order to obtain a paper having a higher air impermeability from a sheet of the above-described polypropylene fibers by the frictional calendering described above, the sheet preferably has a basis weight in the range of from 20 to 400 g/m². The term frictional ratio as used herein denotes the value calculated by the following equation:

$$\text{Frictional Ratio (\%)} = (r_2 - r_1)/r_1 \times 100$$

where r_1 is the peripheral speed of the roll rotating at the lower peripheral speed (m/min), and r_2 is the peripheral speed of the roll rotating at the higher peripheral speed (m/min). Conventional calender rolls composed of an elastic roll and a steel roll may be used as the frictional calender rolls. The only requirement for the calendering rolls is that at least one roll give a friction sufficient to yield the recited frictional ratio to the sheet to be calendered.

Example 1: (a) Into a pressure-resistant tank equipped with a nozzle were charged methylene chloride and polypropylene (5.0% decalin extractable content at 77°C, intrinsic viscosity 1.8, density 0.91). After heating, the value of the nozzle was opened to eject the solution contained in the tank by the action of the vapor pressure of methylene chloride to obtain a microfiber strand of polypropylene.

The strand was then cut into 10 to 15 mm lengths, and then shredded into individual microfibers having an average diameter of 5 μ and an average length of 10 mm. 20% by weight of the resulting polypropylene microfibers and 80% by weight of polypropylene fibers having a birefringence of 3.5 x 10^{-2}, a density of 0.92, a fineness of 2 deniers, and a length of 35 mm were uniformly blended and placed onto a random webber having a stroke of 1,000 mm to produce a nonwoven web having a weight of 18 g/m^2 at a rate of 7 m/min.

The five layers of the resulting webs were pressed under a pressure of 2 kg/cm^2 (gauge pressure) at a temperature of 100°C for 5 seconds, and were needled at a rate of 100 needlings/m^2 using No. 25 felting needles to obtain a uniform sheet having a weight of 90 g/m^2, a thickness of 3 mm and good air impermeability.

The sheet was then continuously subjected to frictional calendering using a universal calender comprising a heated steel roll (surface temperature: 150°C), and a paper roll comprising a plurality of laminated paper discs at a frictional ratio of 70%, a calendering temperature of 155°C, a nip-pressure of 220 kg/cm, and a calendering rate of 3 m/min to obtain the polypropylene paper having the properties shown in the table below.

(b) A polypropylene paper was prepared in the same manner as described in (a) except that the microfibers were prepared from polypropylene having a decalin extractable content of 22% at 77°C, an intrinsic viscosity of 1.5 and a density of 0.90, and 35% by weight of these microfibers were blended to form the nonwoven web. The properties of the resulting paper are shown in the table below.

Example 2: A filament strand was obtained by the melt-spinning method from a polypropylene having an intrinsic viscosity of 2.0, a decalin extractable content of 5% at 77°C and a density of 0.91. The filament strand was then cut into lengths of 5 to 6 mm. The strands thus cut were dispersed in water and were then separated into individual microfibers having an average diameter of 5 μ and an average cut length of 2.5 mm, using a disk refiner.

25% by weight of the above microfibers, 75% by weight of polypropylene fibers having a birefringence of 3.5 x 10^{-2}, a density of 0.92, a fineness of 2 denier and a cut length of 5 mm, and 15 parts by weight (based on the total weight of both fibers) of a polyvinyl alcohol binder were dispersed in water containing a small amount of a nonionic surface active agent to yield a 2% fiber concentration.

To the resulting dispersion there was added an aqueous solution of carboxymethylcellulose having an etherification value of 0.65% (the carboxymethylcellulose having a viscosity of from 150 to 250 cp in a 1% aqueous solution at 20°C) at a level of 100 ppm. The wet sheet was formed from the dispersion at a rate of 150 m/min using a cylinder machine. The surface temperature of the Yankee drier in the papermaking machine was maintained at a temperature of 100°C, which was sufficient to melt the fibrous polyvinyl alcohol binder.

The resulting sheet having a weight of 80 g/m² was then heat-treated at a temperature of 150°C for 1 minute without any tension applied thereto to yield a water-resistant paper with a weight of 110 g/m². The paper thus obtained was immersed in boiling water for about 10 minutes and then washed with water to remove the polyvinyl alcohol, carboxymethylcellulose, surface active agent and other impurities. The paper was subsequently washed with deionized water to remove any ionic substances present, followed by air drying. The sheet was thereafter calendered at a temperature of 155°C, a friction ratio of 50%, a nip-pressure of 200 kg/cm and a calendering rate of 2 m/min using the same calender as was used in Example 1. The polypropylene paper thus obtained was found to have the properties as shown in the following table.

	Example		
	1a	1b	2
Thickness, μ	120	120	120
Air impermeability, G-sec/100 cc	8,000	7,800	1,200
Tensile strength, kg/mm²	6.0	4.8	6.2
Oil resistance (tensile strength after heating in one oil immersion, kg/mm²)	5.7	3.5	5.9
Dissipation factor at 80°C, %	0.04	0.04	0.04

Example 3: A side-by-side bicomponent fiber consisting of 25% polyethylene having a density of 0.962 g/cc and 75% polypropylene having intrinsic viscosity of 2.0 and a density of 0.91 g/cc was drawn at a draw ratio of 400% at a temperature of 100°C and heated under steam pressure of 2.0 kg/cm² in an extended state to yield a 3 denier fiber. The polypropylene portion of the bicomponent fiber thus obtained was found to have a birefringence of 2.9 x 10^{-2}, determined after removing the polyethylene portion. The side-by-side bicomponent fiber treated as above was then cut into lengths of 6 mm using a rotary cutter.

60% by weight of a polypropylene fiber having a fineness of 2 denier, a birefringence of 3.0 x 10^{-2}, a density of 0.92 and a length of 6 mm, 40% by weight of the side-by-side bicomponent fiber described above, and 15 parts by weight (based on the total amount of both fibers) of a fibrous polyvinyl alcohol binder were dispersed in water containing a small amount of a nonionic surface active agent to obtain a fiber dispersion with a 1.5% fiber concentration.

Carboxymethylcellulose (0.65% etherification value, a viscosity of 150 to 200 cp in a 1% aqueous solution at 20°C) was then added to the above fiber dispersion at a concentration of 100 ppm, and a sheet having a weight of 60 g/m² was made by using a cylinder machine. This sheet was subsequently subjected to heat treatment at a temperature of 140°C without any extension to obtain a sheet having a weight of 110 g/m² and displaying excellent mechanical strength upon water immersion. This sheet was washed in boiling water for about 10 minutes in order to remove the polyvinyl alcohol, the carboxymethylcellulose, the surface active agent and washed again with deionized water to remove any ionic substances present.

After this sheet was dried, it was then subjected to frictional calendering at a temperature of 105°C, a nip-pressure of 132 kg/cm and a frictional ratio of 18% to obtain a paper having an air impermeability of 2,000 G-sec/100 cc, a dissipation factor of 0.058% at 80°C (impregnated with liquid paraffin) and a tensile strength of 5.9 kg/mm^2.

REMOVAL OF IONIC SUBSTANCES IN PULP TO BE USED FOR PREPARING ELECTRICAL INSULATING PAPER

The process developed by *H. Fujita, M. Ishitobi, H. Suga, F. Itohara, M. Matsushita and H. Yamamoto; U.S. Patent 3,835,006; September 10, 1974; assigned to Dainichi-Nippon Cables, Ltd. and Tomoegawa Paper Manufacturing Co., Ltd., Japan* comprises continuously passing a pulp slurry prepared by dispersing a pulp in a liquid medium having a specific concentration between a pair of facing voltage-applied electrodes to remove the ionic substances. Preferably, the pulp slurry is passed between a pair of facing voltage-applied electrodes with ion carrier layers interposed between each of the electrodes and the flow layer of pulp slurry. The term ion carrier layers is intended to cover a flowing liquid medium having an electric conductivity of at least 10 μmho/cm.

The pulps which can be treated by the method include all kinds of wood pulp obtained by dissociation in accordance with conventional methods, pulps which are obtained by beating these pulps or treating them otherwise, pulps prepared from synthetic polymers, or mixtures obtained by mixing the above-described pulps with materials such as mica, glass flakes or asbestos. It is necessary that when a pulp slurry is passed between a pair of facing electrodes, the slurry should have an electric conductivity of at least 10 and up to 5,000 μmho/cm. An especially preferred electric conductivity of the pulp slurry is from 50 to 3,000 μmho per centimeter.

The time required to reduce the ash content, i.e., reduce the content of ionic substances from the pulp prepared from the slurry so as to obtain a lower ash content, differs and depends upon the electric conductivity of the pulp slurry, and the concentration of the pulp slurry and the rate of slurry feeding. It usually takes about 1 second to 10 minutes. The pulp is dispersed in an appropriate liquid medium in which an electrolyte can dissociate into ions to prepare a slurry, and then the pulp dispersion is fed between a pair of facing electrodes. A slurry can be alternatively prepared by adding previously loosened pulp into a liquid medium flowing between a pair of facing electrodes.

The use of water as the liquid medium is most economical. Tap water, industrial water, distilled water and deionized water can be used. In order to increase the electric conductivity of the water employed, it is preferred to add an electrolyte which dissociates into ions in a liquid medium such as these types of waters, so long as the electrolyte added does not adversely affect the pulp. Almost any of the water-soluble metal salts and ammonium salts may be added. By adjusting the amount of the salt to be added, the desired electric conductivity of the water can be attained. The most preferred water-soluble metal salts are the alkali metal salts such as the lithium, sodium and potassium salts, the alkaline earth metal salts such as the calcium, magnesium, barium and strontium salts, the ammonium salts of inorganic acids, carboxylic acids and amino acids, and the alkali metal phenolates.

As the liquid medium, a mixture of water with a water-soluble organic material can also be used. The water-soluble organic substances are selected from the group consisting of lower aliphatic alcohols, glycols, glycerins, ethers, carboxylic acid esters, aldehydes, ketones, ketone alcohols, and acid amides. The concentration of pulp in the pulp slurry can range from 0.05 to 6.0% by weight. The pulp concentration is defined by the following equation:

$$\text{Pulp concentration (percent)} = \frac{A}{B} \times 100$$

where A is the dry weight of the pulp (the degree of drying stipulated in JIS P-8127) in the pulp slurry of weight B. The preferred pulp slurry concentration is 0.05 to 5% by weight, an especially preferred concentration being 0.5 to 3% by weight.

A preferred apparatus is shown in Figure 5.1a. The chamber **2** provides a diaphragm **8** by which ion carrier layers **12** and **13** are isolated from the pulp slurry passageway **9**. A liquid medium is introduced from pipes **16** and **17** and discharged from pipes **14** and **15**, and vice versa. The pipes **14** and **16** are extended through the upper wall and pipes **15** and **17** through the bottom wall.

Preferable diaphragms are, for example, woven, nonwoven fabrics and papers which are composed of natural and/or synthetic polymer fibers such as polyethylene, polypropylene, polyamide, polyester or cellulose, in addition to sintered glass plates, and open cellular plastic foamed sheets. Preferably, these diaphragms have the desired mechanical strength when immersed in the liquid and if desired, they may be reinforced with a suitable supporting member. Unglazed plates and semipermeable membranes such as bladders, cellophanes, etc. which are conventionally used in electrodialysis, are of no use as the diaphragm.

FIGURE 5.1: APPARATUS FOR REMOVING IONIC SUBSTANCES FROM PULP

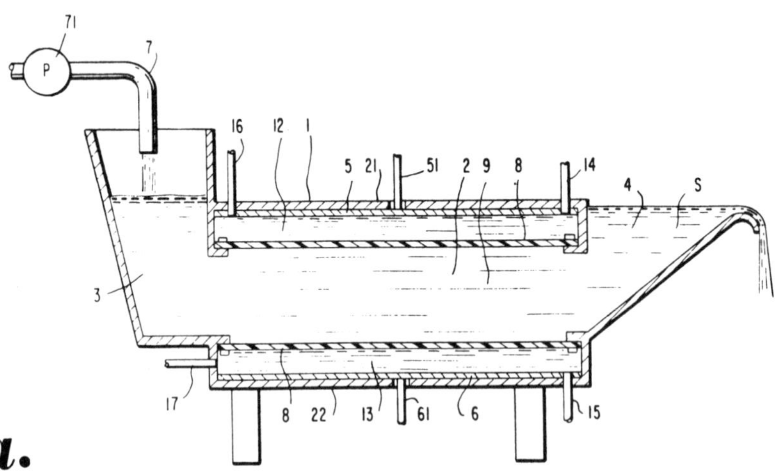

a.

Longitudinal Section of Treating Apparatus Which Includes Diaphragms

(continued)

FIGURE 5.1: (continued)

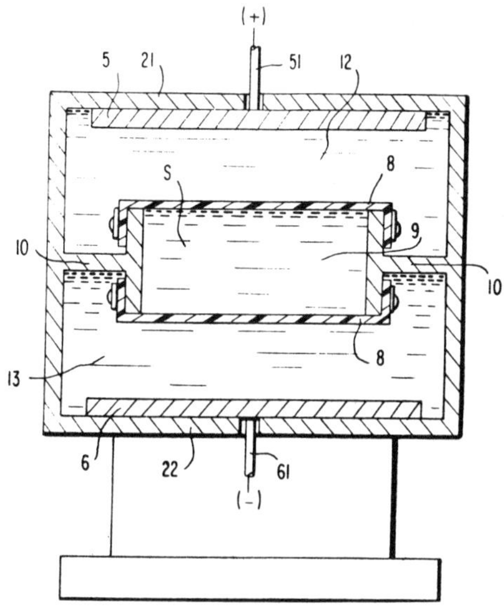

b.

Transverse Sectional View of Treating Chamber with Diaphragm

Source: U.S. Patent 3,835,006

The pulp slurry is passed through the pulp slurry passageway **9** from the chamber **3**, and reaches the chamber **4**, and during the passage through the passageway, undergoes the extraction treatment in the electric field. The distance between diaphragm **8** and each of the electrodes is 0.5 to 50 cm, preferably 1 to 30 cm. The treatment method where the pulp slurry passageway is isolated from the ion carrier layer using diaphragms is most recommended from a practical standpoint and to prevent recontamination of the resultant pulp with ionic substances. Preferable diaphragms have a drainage time, as measured by JIS P-3801 (1956) Item 7.5, of at least 100 seconds, and, when impregnated with a liquid having an electric conductivity of at least 100 μmho/cm have an electric conductivity of at least $1/1000$ of the electric conductivity of the liquid itself.

A type of cross-sectional arrangement of chamber **2** equipped with the diaphragm is shown in Figure 5.1b. The reference numeral **10** represents an electrical insulating plate provided to prevent the extra current flowing outside the passageway, whereby the effective current is concentrated in the region of the passageway. The effective removal of ionic substances is not accomplished in the pulp fiber unless the electric current flows through the pulp slurry passageway. In the examples the pulp sheet used had these properties: Thickness, 3 cm; width, 29 cm; and ash content, 0.40 wt % (measured in accordance with JIS C-2111-9).

The electric conductivities of deionized water, pulp slurry and the liquid medium in the ion carrier layers were measured using a conductivity measuring electrode. This electrode is composed of a pair of platinum wires (a and a') with a diameter of 1 mm, which are separated at a distance of 2 mm by insulating resin (c and c'), its probe being 3 cm in length. This electrode is set in a given position to be measured, and lead wires b and b' are connected to a Wheatstone bridge having an input source of a frequency of 1 kHz, so as to measure the corresponding alternating current conductivity. Its electrode constant was determined utilizing a liquid having a known electric conductivity.

Example 1: Comparative Example Using Unsuitable Diaphragm – A given amount of a pulp slurry was treated without any movement using an apparatus of the type shown in Figure 5.1a. The treating chamber **2** was 1 meter long, 30 cm wide and 5.5 cm high. A platinium plate serving as an anode **5** was attached to the inner surface of the upper wall, and mercury serving as a cathode **6** was settled on the bottom wall in a depth of 0.5 cm (the distance between the electrodes was 5 cm).

This chamber **2** was equipped with the passageway **9** which was constructed in the form of a rectangular tube with a membrane made of cellophane, 0.1 mm thick, being separated 0.5 cm apart from both surfaces of electrodes. Here cellophane as a membrane has a drainage time of above 10^5 sec and specific conductivity index of 3×10^{-4}, as determined as described. The passageway was communicated with the chamber **3** and the chamber **4** at both ends.

The pulp sheet was dispersed in deionized water to prepare a pulp slurry having a pulp concentration of 6.5%, and the resultant slurry was charged into the passageway. Deionized water (electric conductivity 0.1 μmho/cm) was introduced at a flow rate of 2 l/hr from inlets **14** and **15** into a space **13** between the diaphragm **8** and the mercury electrode **6** and a space **12** between the diaphragm and the platinum electrode **5**, and discharged from outlets **16** and **17** respectively. A DC voltage of 400 volts was applied between the platinum anode and the mercury cathode. The pulp treated was then washed thoroughly by dispersion in deionized water (electronic conductivity 0.1 μmho/cm) in an amount of 10 cc/g of the water-freed corresponding dried pulp.

The ash content of the resultant pulp was measured. This experiment was repeatedly conducted for varying periods of the voltage application on the virgin pulp slurry. The results obtained are given below.

Time of Applying Voltage, min	Ash Content, %
5	0.36
10	0.34
20	0.31
30	0.29
60	0.26

Example 2: Runs Under Different Conditions – The same apparatus as described in Example 1 was employed for 20 runs, except that a diaphragm of an aromatic polyamide paper (Nomex) was substituted in place of the cellophane. Pulp slurries of various concentrations were prepared with various liquid media from the pulp sheet. Each pulp slurry was fed continuously into the passageway separated with the diaphragm and treated under varying conditions such as time, current, etc.

A DC voltage of 400 v, except for 200 v for 2 runs, was applied between the anode **5** and the cathode **6**; current in the stationary state of the pulp treatment varied between 9 and 190 amp. The electric conductivity of both the ion carrier layers was measured before and during treatment at the specified position which was quite close to the diaphragm **8** and was 50 cm apart from the inlet facing electrodes. In each run, the flowing direction of the liquid medium in the ion carrier layer was the same at both sides. When the flowing direction of the liquid medium in both carrier layers was the same as that of the pulp slurry, it was designated "same." About half the runs were "same;" the remainder were reverse, "opposite." The pulp slurry treated was washed with water in the same manner as described in Example 1.

In addition, in 5 cases a 100-μ-thick paper was made of the corresponding resultant pulp using a hand sheet machine with ionized water. The apparent density of the paper was about 0.51 g/cm^3 and its dielectric dissipation factor in an oil-impregnated state at 100°C was about 0.11%. A 100-μ-thick paper formed from the original pulp sheet using the same hand sheet machine had an apparent density of 0.49 g/cm^3, and a dielectric dissipation factor, in an oil-impregnated state, of 0.16% at 100°C.

The dielectric dissipation factor of the oil-impregnated paper was measured by inserting three sheets of the paper into a flat circular electrode at a temperature of 100°C and a field strength of 10 kv/mm generated by a high-voltage Schering bridge. The paper to be measured was previously dried at 120°C for 4 hours in vacuo at $^1/_{1000}$ mm Hg, and impregnated with a degassed cleaned dodecylbenzene oil.

The treatment included water-washing the treated pulp slurry within 1 to 2 minutes and achieved sufficient removal of ionic substances within very short periods of time, as compared with the case in which no diaphragm was involved.

This process is excellent for production of pulp to use for paper having a low dielectric dissipation factor. However, if the pulp slurry concentration is as high as 6%, it is a little difficult to keep the pulp slurry flowing constantly and small air bubbles can be observed in the pulp slurry.

ELECTROCONDUCTIVE PAPERS

PAPER FOR ELECTROSTATIC PRINTING

Dimethyldiallyl Ammonium Chloride Polymers

Where electrically conductive paper is to be used for nonimpact printing, a substrate, backing, impregnation coating, or layer of electrically conductive material is usually constructed. Various types of nonimpact printing processes are known as electrostatographic, electrophotographic, electrographic, Electrofax, and other processes.

As a rule such processes call for the placement of an electric charge on the paper, which may be accomplished by a corona discharge, for example. The charge is, in most such processes, placed on the paper in darkness. The paper may also contain a photoresponsive or photoconductive layer or material, now popularly a specially treated zinc oxide, which causes the charge to be dissipated in an area where light strikes it, thus leaving a pattern of the charged areas which is a reproduction of the image desired.

The charged area attracts a powdered or other usually particulated image-forming material which may be fused or otherwise treated to make the image permanent. Other processes differ in that the image is created by electrical dissipation of the static charge in nonimage areas; in this and other processes, the common characteristic is an electrically conductive base paper.

J.E. Boothe and M.F. Hoover; U.S. Patent 3,544,318; December 1, 1970; assigned to Calgon Corporation have found the class of water-soluble polymers containing at least about 90% repeating groups of the formula

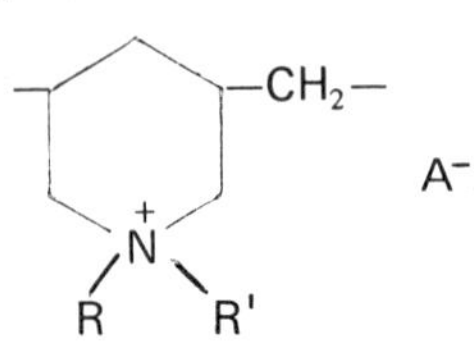

where R is an alkyl group containing 1 to 18 carbon atoms, R' is selected from the group consisting of R and

$$-CH_2-CH_2-\overset{\overset{\displaystyle O}{\|}}{C}-NH_2$$

and A is an anion, may be used in making electroconductive paper.

The preferred homopolymer of the described class is derived from free radical polymerization of dimethyldiallyl ammonium chloride. The polymerization method described by Butler and Angelo, *JACS,* 79, 3128 (1957), may be used satisfactorily to make polymers. Also useful is the homopolymer of the monomer diallylmethyl β-propionamido ammonium chloride.

Diallyl ammonium monomers undergo cyclization during free radical polymerization to form piperidinium rings. The homopolymer of dimethyldiallyl ammonium chloride may be described as a polydimethyl-3,5-methylene piperidinium salt. The homopolymer of diallylmethyl β-propionamido ammonium chloride, for example, is known as poly[N-methyl-N-(β-propionamido)-3,5-methylene piperidinium chloride].

Any copolymerizable monomer which will form a water-soluble polymer with the diallyl monomer may be used for the remaining 10%. Among the monoethylenically unsaturated copolymerizable monomers suitable for use are acrylamide, diacetone acrylamide, N-vinyl pyrrolidinone, and N-methylolacrylamide.

The homopolymers and copolymers may include in their polymerized structures certain compatible polyethylenic unsaturated compounds. Such polyethylenic unsaturated compounds, when present in small amounts (i.e., up to 10%) during the polymerization of a dialkyldiallyl ammonium halide monomer, cause the resulting polymer to be highly branched and/or crosslinked. The branched polymer also imparts superior barrier properties to the paper substrate preventing solvent from the application of the photoconductive upper layer from diffusing into the paper. The preferred branching agents are methylenebisacrylamide, triallylamine hydrohalide salts, triallylquaternary ammonium salts having as a fourth group an alkyl group of 1 to 4 carbon atoms, and tetraallyl ammonium salts.

Solutions of the polymers may be applied to paper or cellulosic webs by coating, dipping, brushing, wet end addition, etc. Concentrations of the polymer may vary widely to fit the method of handling. Generally about 0.5 to 3.0 pounds of polymer should be applied per 3,000 square feet but for some purposes as little as 0.1 pound will provide sufficient conductivity.

Conductivity measurements for conductive coatings on paper may be made on the conductive areas only; that is, the electrodes of the conductivity device may be simply attached to the conductive surface of the paper. Generally speaking, papers adapted for use in various types of nonimpact printing may have surface resistivities in the range of about 2.5×10^5 to about 3.0×10^9.

Example 1: Use of Different Polymers — (a) A 4% by weight aqueous solution of a copolymer of dimethyldiallyl ammonium chloride and methyldodecyldiallyl ammonium chloride in the molar ratio of 17.5 to 1 was applied to Bergstrom Paper Company's Indsen No. 1 raw coating stock (basis weight 46 lb per 3,300 square feet) using a 3 mil, wire-wound Meyer rod. This rod applies approximately 0.3 lb of pure polymer per 3,000 square feet of paper. The sample was then dried in an oven at 105°C for 20 to 30 minutes.

An 8 inch circular sample was cut and placed into a constant conditions room for 16 hours at 50% relative humidity and 72°F. After this period, the sample was conditioned for 72 hours at 23% relative humidity. Current measurements were made after applying 100 volts DC using Keithley equipment consisting of a Model 240 Regulated High Voltage Supply, a Model 6105 Resistivity Adapter and a Model 610-B Electrometer. The resistivity adapter is suitable for determining both surface and volume resistivities.

Electrodes are constructed of stainless steel; however, for measurements at relative humidities of greater than 70% a special graphite center electrode was used. Current readings were converted to obtain resistivity values. The surface resistivity of this polymer was 5.9×10^8 ohms/square.

(b) A 4% by weight aqueous solution of poly(methyl β-propionamidodiallyl ammonium chloride) was applied to the test paper as above. After treatment as in (a), this paper gave a surface resistivity value of 2.0×10^{10}.

(c) A 4% by weight solution of a copolymer of dimethyldiallyl ammonium chloride (93% by weight) acrylamide (4%) and acryonitrile (3%) was applied and treated as in (a). At a relative humidity of 12%, the surface resistivity was 8.6×10^{10}.

(d) A 4% solution of homopolymer of diethyldiallyl ammonium chloride was applied and treated as in (a). The surface resistivity at 20% relative humidity was 7.9×10^{11}.

(e) A 4% by weight solution of a copolymer of dimethyldiallyl ammonium chloride copolymerized with 0.03 weight percent methylene bisacrylamide was applied and treated as in (a). The surface resistivity at 23% relative humidity was 1.8×10^{10}.

Example 2: Conductivity at Different Humidity — Poly(N,N-dimethyl-3,5-methylene piperidinium chloride) was prepared in aqueous solution and applied to a paper by a Time-Life bench coater. After drying, conductivity measurements showed the following surface resistances at the relative humidities indicated in terms of ohms/square:

Relative Humidity	- - - - - Surface Resistivity - - - - -	
	(a)	(b)
20	3×10^9	6×10^8
45	6×10^7	5×10^7
80	7×10^5	6×10^5

The polymer was used in a concentration of two pounds for (a) and three pounds for (b) per 3,000 square feet.

Polymers of Amine Salt Quaternized with Epichlorohydrin

S.N. Lewis, R.F. Merritt and W.D. Emmons; U.S. Patent 3,702,799; Nov. 14, 1972; assigned to Rohm and Haas Company are concerned with monoethylenically unsaturated acid esters monomers, such as those of acrylic or methacrylic acid, containing a quaternary ammonium halide group having a substituent on the quaternary nitrogen atom of the formula $-CH_2CH(OH)CH_2X$ wherein X is chlorine, bromine or iodine.

The process also pertains to addition polymers of such monomers, and methods of producing the monomers and polymers. The monomers and polymers are stable at acid pH values, although capable of reacting with other reactants for the hydroxyl and/or halide group. On heating, they are self-reactive and become crosslinked. At alkaline pH values, they become self-reactive and the polymers crosslink themselves by slow reaction even at room temperature. However, the crosslinking reaction becomes quite rapid on heating. The polymers are used as wet-strength agents in making paper and as electroconductive aids in the making of various electroconductive papers for electrostatic image reproduction systems.

Reaction of the amine salt polymer (whether homopolymer or copolymer) with the epihalohydrin provides a polymer having units containing quaternary ammonium groups of the formula

(1)

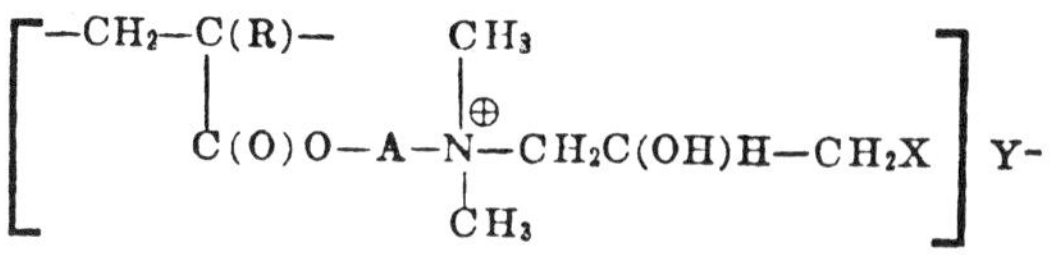

along with some units of the formula

(2)

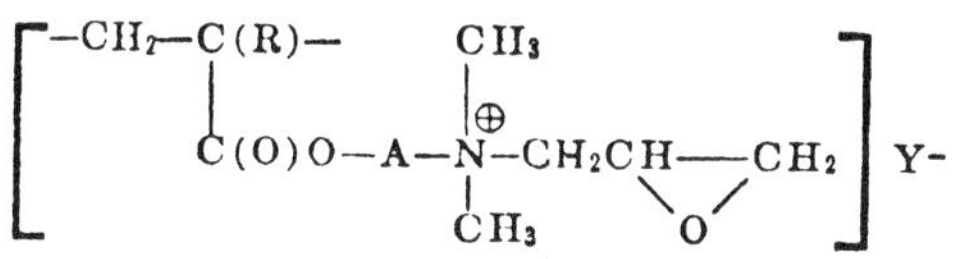

Copolymers of (a) about 0.5 to 30% by weight or more of a monomer of Formula (1) and/or Formula (2) with (b) methacryloxyethyltrimethyl ammonium chloride or acryloxyethyltrimethyl ammonium chloride are of special value in providing paper with higher conductivity for use in specialty applications such as electrostatic copy paper, electrographic printing, and other applications where paper with lower than normal electrical resistivity is required. These water-soluble linear copolymers of this process may have molecular weights in the range of about 20,000 to 2,000,000 or more and may be made by the direct polymerization of the quaternary monomers of types (a) and (b) or their salts.

Alternatively, a homopolymer of dimethylaminoethyl acrylate or methacrylate may be made and this amine-containing polymer may be simultaneously quaternized with the components of a mixture of an epihalohydrin and a methyl halide, such as a mixture of epichlorohydrin and methyl chloride, in the proportions to provide 0.5 to 30% by weight of (a) and the balance of component (b) of the copolymer.

When polymers in the lower molecular weight range are desired, the polymerization of the amine, its salt, or its quaternary may be effected at elevated temperatures, e.g., 40° to 60°C or higher using conventional initiator systems, such as ammonium persulfate with or without sodium hydrosulfite. The higher molecular weight polymers may be obtained at lower temperatures, such as 5° to 10°C and using concentrations of 60% or more with dilution to facilitate handling as polymerization progresses.

To prepare an electroconductive paper, the copolymers of (a) and (b) hereinabove defined may be applied to the paper, or cellulosic web, by the conventional methods used for that purpose, e.g., coating, dipping, brushing, or by wet end addition, etc. The paper used may have a basis weight of 30 to 75 pounds, preferably 30 to 55 pounds, per 3,000 square feet.

The amount of polymer applied to the paper will generally vary within the range of about 1 to 8% (weight) pick-up, depending upon the particular polymer and paper combination used and the degree of electroconductivity which is desired. In some cases, still less might be used. There seems to be no operative upper limit to the amount of polymer applied, except to the extent this is determined by economics. The overall range of from about 1 to 8% pick-up (by weight) is simply a statement of the required amount of polymer to confer electroconductivity properties to the cellulosic web substrate which will adapt it to most commercial uses mentioned above.

As compared to papers made with the water-soluble quaternary ammonium polymers heretofore recommended for improving the electroconductivity of electrostatic reproductive paper, the papers made with the copolymers of this process containing 0.5 to 30 weight percent of units of Formulas (1) and/or (2) with the balance to make 100% of (meth)acryloxyethyltrimethylammonium chloride units have the advantage of retaining their electroconductivity for longer periods of time. This is apparently attributable to the fact that on drying the paper after impregnation or coating thereof with these copolymers, they are crosslinked to a water-insoluble condition.

The fact that these copolymers are somewhat crosslinked on drying also provides an additional advantage of reducing the tendency to block or eliminating it entirely so that the coated paper carrying the electroconductive material can be rolled on itself without fear of damage when it is unrolled for later use.

Also, the crosslinked condition reduces the sensitivity to organic solvents and aqueous media so that the topcoating layers of the photoconductive material such as zinc oxide in a binder dissolved or dispersed in a solvent or in an aqueous medium does not cause any appreciable leaching of the electroconductive agent from the paper at the time of application of the topcoating. Examples of such topcoating layers include styrene-butadiene latexes, solutions in an organic solvent or dispersions in water of a polymer containing from 3 to 50% by weight of acid such as a copolymer of styrene and maleic anhydride, a copolymer of 95% butyl acrylate and 5% itaconic acid or of 70% ethyl acrylate and 30% acrylic acid or methacrylic acid, partially hydrolyzed polyvinyl acetate and alkali-soluble cellulose esters such as carboxyethyl cellulose as well as water-soluble cellulose ethers such as hydroxyethyl cellulose. Such topcoating layers contain zinc oxide as well as one of the binders just mentioned and may contain a dye sensitizer.

Example 1: Monomer Preparation — A flask fitted with a thermometer, stirrer and an addition funnel is charged with 2,000 grams of deionized H_2O and 962 grams of 37% HCl. A supply of dimethylaminoethyl methacrylate (1,570 grams) is gradually added and the temperature maintained at 25°C with external cooling. The epichlorohydrin (1,000 grams) is added all at once and the temperature maintained at 25° to 30°C for 8 hours. Amine titration of this solution confirms the quaternization of the amine function. The 51.6% solution of quaternary monomer (Br No. = 31.0, calculated 28) is essentially entirely (99%) in chlorohydrin form (thiosulfate/oxirane <1%) and is utilized as a solution.

Example 2: (a) Polymer Preparation — A mixture of 62.8 grams of water, 27.0 grams of methacryloxyethyl trimethyl ammonium chloride, and 5.2 grams of the 51.6% solution of the quaternary monomer obtained in Example 1 is purged with nitrogen, heated to 65°C and then 1.0 gram of a 0.1% solution of $FeSO_4 \cdot 7H_2O$ in water, 0.3 gram of ammonium persulfate, and 0.3 gram of sodium hydrosulfite are added. After one hour, the mixture is heated to 85°C and 0.05 gram of ammonium persulfate and 0.05 gram of sodium hydrosulfite is added After one hour, the mixture is cooled to provide a clear solution, 30.2% solids, Brookfield viscosity of 480 centipoises and pH of 1.5.

(b) Use — One part of the polymer of part (a), activated with aqueous caustic equivalent to the amount of structure (1) incorporated into the polymer, is mixed with one part of a hydroxyethylated starch available commercially (Pen-Gum-280) and one part of kaolinite clay in a formulation of 30% solids. This formulation is applied at various levels to a paper base stock (a 40 pound per 3,000 square foot paper made from bleached pulp beaten to a Canadian Standard Freeness of 200 ml) typical of that used for Electrofax copy paper by means of wire wound rods, and dried for one minute at 180°F. Surface resistivity at 15% relative humidity is measured using ASTM Method D256-61. The results are summarized below.

Conductivity Aid	Conductivity Aid, lb/3,000 ft^2 Paper,	Surface Resistivity, ohms/square
Above copolymer	0.5	1.3×10^{11}
Above copolymer	1.0	5.0×10^{10}
Above copolymer	2.5	1.4×10^{10}
Above copolymer	5.0	7.6×10^{9}
Commercial product	0.5	8×10^{10}
Commercial product	1.0	2.4×10^{10}
Commercial product	2.5	5.9×10^{9}
Commercial product	5.0	3.3×10^{9}
None (untreated paper)	-	1×10^{14}

Quaternized Polyethylenimine

The process of *G. Cavagna and W.C. Walker; U.S. Patent 3,479,215; Nov. 18, 1969; assigned to Westvaco Corporation* relates to paper having an electroconductive surface adapted for electrographic printing and containing a polymer comprised of quaternized polyethylenimine. Paper coated with quaternized polyethylenimine has a surface resistivity ranging from about 1×10^8 to 1×10^5 ohms per square centimeter over a relative humidity range of about 8 to 75%.

The molecular weight of the polyethylenimine before quaternization has little effect on the conductive properties of the polymer after quaternization, but molecular weight does affect the viscosity of the coating containing the quaternized polyethylenimine. Thus, the choice of the molecular weight of the material to be used is determined by the desired final viscosity. Polyethylenimines having molecular weights ranging from 1,200 to 100,000, and continuous films of coatings containing these polymers have been applied to paper without any operational difficulties.

The polyethylenimine is preferably reacted with an alkylating agent such as methyl chloride by stirring at a temperature between 40° to 80°C. The pH of the reaction medium, which tends to drop, is preferably kept on the alkaline side, between about 8 and 9, in order to achieve higher yields of the quaternized polymer. The reaction is complete after several hours and the product can be used in solution, as it is, or it can be dried. If the quaternized polymer is used in solution, the pH can be adjusted towards neutral with an acid, such as hydrochloric acid. If the quaternized polymer is dried before use, it can be neutralized either before drying or before its use when it is put into solution. The alkaline product performs satisfactorily as an electroconductive coating, but the polymer, neutralized after quaternization, exhibits even higher conductive properties.

Instead of maintaining the reaction medium at a pH between about 8 and 9, as mentioned above, the reaction between polyethylenimine and the alkylating agent may be allowed to proceed without pH adjustment. Under these conditions, pH drops from the natural alkaline level of the polyethylenimine to about neutral, and quaternization of at least about 50% of the available nitrogen atoms in the polymer can be achieved. In any event, regardless of whether or not the pH is adjusted during the reaction, the number of quaternary ammonium groups present should be at least about 100 for every 1,000 atoms forming the polymeric chain in order that the quaternized polyethylenimine possesses satisfactory conductive properties.

In preparing paper having an electroconductive surface containing the electroconductive polymers, the polymers can be applied as a continuous film to paper by coating, spraying, dipping, brushing or by any other suitable application. The quaternized polymer is preferably applied to paper, as a coating, in the form of an aqueous dispersion or solution. The coated paper is then dried to produce a product having an electroconductive surface comprised of a continuous film of the quaternized polyethylenimine. Standard coating devices, such as a blade coater, have been used, and the coated paper has been dried by the use of radiant energy from heat lamps. The electroconductive polymer has been applied at coat weights ranging from about 0.5 to 29.0 pounds of coating per ream (500 sheets, 25 x 38 inches), with a preferred coat weight being about two pounds per ream.

Other polymers may be mixed with quaternized polyethylenimine before application of the coating to a paper substrate. A polymer, such as starch may be used as a component of the coating formulation to prevent strike-through of the electroconductive polymer by keeping the latter on the surface of the paper. The starch quaternized polyethylenimine coating may contain any proportion of starch up to about a weight ratio of about 1 to 1 with the electroconductive polymer. The coated paper is tested for surface resistivity by a procedure substantially like that described in "Standard Methods of Test for Insulation Resistance of Electrical Insulating Materials," ASTM D-257-46.

In the actual practice of preparing a printing surface for an electrographic printing process, a coating of a photoconductive material, such as zinc oxide, is applied over the electroconductive surface containing the electroconductive polymers.

Example 1: Preparation of Quaternized Polyethylenimine — In a reaction vessel provided with a stirrer, thermometer, and condenser, 32.25 parts by weight of dry polyethylenimine, having a molecular weight of about 1,200, were mixed with 600 parts by weight of water. The temperature of the solution was raised to 60°C, and then methyl chloride was bubbled through the solution. Throughout the quaternization reaction, the pH, which tends to drop, was maintained between about 8 and 9 by adding aqueous sodium hydroxide. After about 80 parts by weight of methyl chloride were added to the solution, the quaternization reaction was allowed to continue for about ten hours. The reaction mixture was then freeze dried by using the well-known Virtis Unitrap apparatus, and about 90 parts by weight of dry quaternized polyethylenimine was obtained.

Example 2: Use of Alkaline Product — (a) Part of the dried quaternized polyethylenimino of Example 1 was dissolved in water to form a 30% solids coating. A sheet of bleached sulfate paper having a basis weight of about 50 pounds per ream (500 sheets, 25 x 38 inches), was coated on one side with a continuous film of the coating, using a No. 26 Meyer rod to produce a coat weight of about 8.3 pounds per ream. The coated paper was dried by radiant energy, and then four test samples were cut from the polymer coated paper.

These test samples were conditioned in air having a relative humidity of about 8%, at a temperature of about 25°C, for about 24 hours. The test samples were then tested for surface resistivity, and the samples exhibited an average surface resistivity of about 4.3×10^8 ohms per square centimeter. The polymer coated paper had an electroconductive surface adapted for electrographic printing.

(b) The procedure was repeated, and the test samples, which had a coat weight of about 22.5 pounds per ream, were tested after being conditioned in air having a relative humidity of about 75%. The average surface resistivity of the test samples was about 3.1×10^5 ohms per square centimeter.

Example 3: Use of Neutralized Product — (a) The procedure of Example 1 was repeated, except that the 30% solids coating comprising quaternized polyethylenimine was neutralized to pH 6.8 with hydrochloric acid before the coating was applied to the paper. The paper was then coated, as in Example 2 with about 6.0 pounds per ream of the coating, dried, and four test samples were cut from the coated paper and conditioned similarly as in Example 2. The samples had an average resistivity of about 1.2×10^8 ohms per square centimeter. The polymer coated paper had an electroconductive surface adapted for electrographic printing.

(b) The procedure was repeated, and the test samples, which had a coat weight of about 12.5 pounds per ream, were tested after being conditioned in air having a relative humidity of about 75%. The average surface resistivity of the test samples was about 3.1×10^5 ohms per square centimeter.

Example 4: Use of Starch-Quaternized Polyethylenimine Derivative — (a) Example 1 was repeated to obtain the alkaline product. Cornstarch was gelled by cooking a 15% by weight starch-water mixture for 15 minutes at 82°C. Then a coating was prepared comprising a mixture of 50% by weight of the gelled starch and 50% by weight of a 15% solution in water of the quaternized polyethylenimine.

The starch-quaternized polyethylenimine coating was applied to paper at a coat weight of about 5.2 pounds per ream (about 2.6 pounds per ream of the electroconductive polymer), and the coated paper was dried, conditioned at 8% relative humidity, and tested as in Example 2. The average surface resistivity of the test samples was about 1.8×10^9 ohms per square centimeter. The coated paper met the standards of the industry for use in the electrographic printing processes.

(b) The procedure was repeated, and the test samples, which were coated with about 7.4 pounds per ream of the starch-quaternized polyethylenimine coating (about 3.7 pounds per ream of the electroconductive polymer), were tested after being conditioned in air having a relative humidity of about 75%. The average surface resistivity of the test samples was about 7.0×10^5 ohms per square centimeter.

Example 5: Use of High Molecular Weight Polyethylenimine — The procedure of Example 1 was repeated, except that the polyethylenimine employed had a molecular weight of about 40,000 to 60,000. The test samples were coated with about 4.5 pounds per ream of the coating, dried, conditioned in air having a relative humidity of about 75%, and then tested as in Example 2. The average surface resistivity of the test samples was about 5.0×10^5 ohms per square centimeter.

Unsaturated Quaternary Monomers and Polymers of Methacrylic Acid

The process of *H.J. Cenci and T.E. Stevens; U.S. Patent 3,689,468; Sept. 5, 1972; assigned to Rohm and Haas Company* is concerned with monoethylenically unsaturated acid ester monomers, such as those of acrylic or methacrylic acid, containing a bis-quaternary ammonium halide group, such as one of the general formula

$$H_2C{=}C(R){-}C(O){-}O{-}A{-}\overset{R^\circ}{\underset{R^1}{\overset{+}{N}}}{-}CH_2{-}CH{=}CH{-}CH_2{-}\overset{R^2}{\underset{R^3}{\overset{+}{N}}}{-}R^4 \quad X^- \quad Y^-$$

R being H or methyl,
A being C_2-C_6 alkylene,
R° being lower alkyl,
R^1 being lower alkyl,
R^2 being lower alkyl,
R^3 being lower alkyl,
R^4 being lower alkyl,
X being an anion,
Y being an anion.

It is also concerned with addition polymers of such monomers. The monomers and polymers are reasonably stable at both alkaline and acid pH values under normal temperature and pressure conditions although capable of reacting with other reactants at the point of unsaturation in the alkenyl linkage between the two quaternary nitrogen groups. These monomers and polymers find use as antistatic agents for textiles and other articles formed of hydrophobic materials, as wet-strength agents in making paper and as electroconductive aids in making of various electroconductive papers for electrostatic image reproduction systems. To prepare an electroconductive paper, homopolymers and copolymers may be applied to the paper, or cellulosic web, by the conventional methods used for the purpose, e.g., coating, dipping, brushing, or by wet end addition, etc. The paper used may have a basis weight of 30 to 75 pounds, preferably 30 to 55 pounds per 3,000 square feet. The amount of polymer applied to the paper will generally vary within the range of about 1 to 8% (weight) pick-up, depending upon the particular polymer and paper combination used and the degree of electroconductivity which is desired.

As compared to papers made with the water-soluble quaternary ammonium polymers heretofore recommended for improving the electroconductivity of electrostatic reproductive paper, the papers made with certain of the copolymers of this process, especially the copolymers containing 0.5 to 30 weight percent of units of the above general formula, 0.5 to 20 weight percent of a reactive monomer such as an N-methylolacrylamide or a monomer containing a glycidyl group, such as glycidyl methacrylate or one of the quaternary ammonium monomers having a glycidyl group attached to the quaternary nitrogen atom, with the balance to make 100% of (meth)acryloxyethyltrimethylammonium chloride units have the advantage of retaining their electroconductivity for longer periods of time. This is apparently attributable to the fact that on drying the paper after impregnation or coating thereof with these copolymers, they are crosslinked to a water-insoluble condition and/or reacted with the cellulose of the paper.

The fact that these copolymers containing reactive groups are somewhat crosslinked and/or reacted with the paper on drying also provides an additional advantage of reducing the tendency to block or eliminating it entirely so that the coated paper carrying the electroconductive material can be rolled on itself without fear of damage when it is unrolled for later use.

Also, the crosslinked and/or reacted condition reduces the sensitivity to organic solvents and aqueous media so that the topcoating layers of the photoconductive material such as zinc oxide in a binder dissolved or dispersed in a solvent or in an aqueous medium do not cause any appreciable leaching of the electroconductive agent from the paper at the time of application of the topcoating.

Example 1: Polymer Preparation — (a) A 2 liter glass polymerization flask is charged with 1,000 grams of deionized water and sparged with nitrogen for one hour. The solution is cooled to 15°C and 138.5 grams of 37% aqueous hydrochloric acid is added, followed by 220 grams of dimethylaminoethyl methacrylate. The temperature rises to 30°C and the mixture is further warmed to 32°C. Then, 10 ml of a 0.1 N solution of $FeSO_4 \cdot 7H_2O$, 1.5 grams of ammonium persulfate, and 1.5 grams of sodium hydrosulfite are added. Forty-five minutes after the initial catalyst addition, 0.2 gram of ammonium persulfate and 0.2 gram of sodium hydrosulfite are added. Charges of 0.1 gram of ammonium persulfate and 0.1 gram of sodium hydrosulfite are added 30 minutes later, followed by 0.05 gram of ammonium sulfate and 0.05 gram of sodium hydrosulfite 30 minutes later. The polymer solution at 25°C has a solids content of 21.7%, a Brookfield viscosity of 55 centipoises, and an amine content of 1.06 milliequivalents per gram.

(b) A 200 gram portion of this polymer solution and 36 grams of water are stirred in a glass vessel, and the pH is adjusted to 7.2 with 25% aqueous sodium hydroxide while the solution is warmed to 50°C. Then, 39.0 grams of 4-chloro-2-butenyltrimethylammonium chloride is added in one portion. The reaction temperature is maintained at 50°C and the pH of the reaction mixture is maintained at 6.8 to 7.1 by the periodic addition of 25% aqueous sodium hydroxide.

At the end of two hours, titration indicates that 87% of the amine has been quaternized; another 3.9 grams of 4-chloro-2-butenyltrimethylammonium chloride is then added. After one hour at 50°C and 16 hours at ambient temperature, quaternization is complete as indicated by a lack of amine functionality upon acid-base titration. The final polymer solution has a solids content of 32.4% and a Brookfield viscosity of 117 centipoises.

(c) To prepare α-(α'-chloro-p-xylylene)trimethylammonium chloride 34 grams of trimethylamine is added to 100 grams of p-xylene dichloride in 200 grams of acetone. The product, 137 grams, is isolated by filtration.

(d) To 200 grams of the poly(dimethylaminoethylmethacrylate hydrochloride) prepared in (a) above there is added 25% aqueous sodium hydroxide until the pH rises to 7.0. The solution is warmed to 60°C and 36.9 grams (0.77 equivalent) of the α-(α'-chloro-p-xylylene)trimethylammonium chloride is added. The pH of the solution is maintained at 7.0 by the addition of 25% aqueous sodium hydroxide.

After 70 minutes, quaternization of the amine is 77% complete; 13 grams (0.23 equivalent) of the above ammonium salt is added, the pH adjusted to 8.5, and the reaction mixture set aside for 3 days. The finished polymer solution, whose pH is adjusted to 6.0 with hydrochloric acid, has a solids content of 33.2% and a Brookfield viscosity of 157.7 centipoises.

(e) The polymers of (b) and (d) are tested as conductivity aids for electroconductive paper in the following way and the results are reported below. One part of the polymer to be tested, one part of a hydroxyethylated starch available commercially (Pen Gum 280), and one part of kaolinite clay are mixed to produce a 30% solids mixture. This formulation is applied at various levels to a paper base stock (40 pounds per 3,000 square feet paper made from bleached pulp beaten to a Canadian Standard Freeness of 200 ml) typical of that used for Electrofax copy paper by means of wire wound rods, and dried for one minute at 180°F. Surface resistivity at 15% relative humidity is measured using ASTM Method D256-61.

Conductivity Aid	Conductivity Aid, lb/3,000 ft^2 Paper	Surface Resistivity, ohms/square
Polymer of (d)	1.5	4.8×10^{10}
Polymer of (d)	2.0	4.2×10^{10}
Polymer of (d)	2.75	3.5×10^{10}
Polymer of (b)	1.5	6.5×10^{9}
Polymer of (b)	2.0	5.2×10^{9}
Polymer of (b)	2.75	3.9×10^{9}
Commercially available product	1.5	4.9×10^{9}
Commercially available product	2.0	4.3×10^{9}
Commercially available product	2.75	3.6×10^{9}
None (untreated paper)	–	1×10^{14}
Second commercial product	1.5	2.6×10^{9}
Second commercial product	2.75	1.4×10^{9}

The following examples are taken from a process by *B.B. Kine and W.D. Emmons; U.S. Patent 3,766,156; October 16, 1973; assigned to Rohm and Hass Company.*

Example 2: (a) A solution (100.0 grams) in water containing 22 grams of the hydrochloride of dimethylaminoethyl methacrylate is mixed with 20.5 grams of (2,3-epoxypropyl)trimethylammonium chloride. The mixture is heated at 50°C for 5 hours at a pH of 7.0 to 7.5 and allowed to stand for 24 hours at 25°C. Quaternization is essentially complete. The solution is adjusted to pH 3 with 37% hydrochloric acid. The resulting monomer of the formula

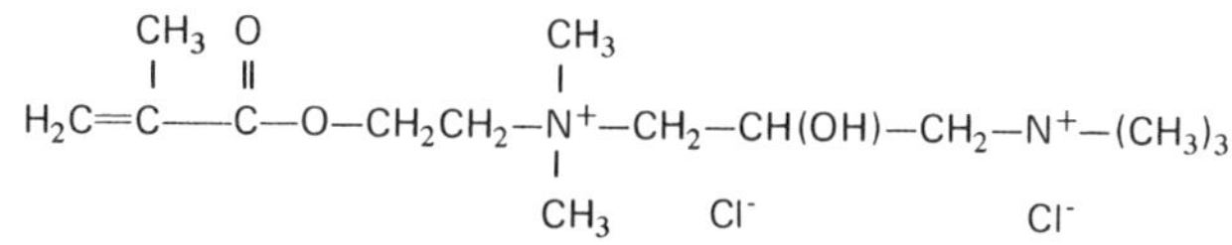

is polymerized by the addition, at 65°C, of 1 ml of a 0.1% solution of $FeSO_4 \cdot 7H_2O$, 0.2 gram of ammonium persulfate and 0.3 gram of sodium hydrosulfite. The temperature is maintained at 65°C for one hour, then raised to 85°C and 0.05 gram of ammonium persulfate and 0.05 gram of sodium hydrosulfite are added. After one hour at 85°C the solution is cooled to 25°C to provide a polymeric solution of 31.6% solids.

(b) The procedures of (a) are repeated six times except that in each instance the amine is replaced by a molar equivalent amount of a different one of the following amines:

3-(dimethylamino)propyl acrylate
2-(dimethylamino)propyl methacrylate
2-(dimethylamino)hexyl methacrylate
4-(dimethylamino)butyl acrylate
2-(diethylamino)ethyl acrylate
2-(dibutylamino)ethyl methacrylate

(c) Each of the polymers above is formulated with hydroxyethylated starch and kaolinite clay and applied to paper in the following manner. One part of the respective polymer, one part of a hydroxyethylated starch available commercially (Pen Gum 280), and one part of kaolinite clay are mixed to produce a 30% solids mixture. This formulation is applied at various levels from 1.5 pounds to 2.75 pounds per 3,000 square feet to a paper base stock (40 pounds per 3,000 square feet paper made from bleached pulp beaten to a Canadian Standard Freeness of 200 ml) typical of that used for Electrofax copy paper by means of wire wound rods, and dried for one minute at 180°F. The resulting papers show reduced resistivity and can be used as electrostatic reproductive papers.

(d) The papers obtained are given a dielectric coating layer by topcoating with the following formulation:

Material	Parts
Ammonium salt of a copolymer of 50% butyl acrylate, 25% methyl methacrylate, 10% styrene, and 15% methacrylic acid (35% aqueous solution)	57.2
Copolymer of 32% ethyl acrylate and 68% methacrylic acid acid (20% aqueous dispersion)	1.25
Photoconductive ZnO (Photox 801)	140
Bromphenol Blue in methanol (½% solution)	5.6
Ammonium hydroxide (28%)	0.8
Water	27.1

This coating is applied at 15 pounds per 3,000 square feet and air-dried. The polymers containing the quaternary ammonium groups show little or no migration into the dielectric coating and good images are produced.

Example 3: A slurry of 19.7 grams of calcium hydroxide in 100 grams of water is added to 500 grams of the polymer solution obtained in Example 1(a) above. The solution is stirred for 6 hours under a positive pressure (30 mm of Hg) of methyl chloride. At this time, 10 grams of (2,3-epoxypropyl)trimethylammonium chloride is added and the mixture is stirred another 16 hours. Titration of amine functionality indicates that quaternization is complete. The pH of the solution is adjusted to 1.2 with aqueous hydrochloric acid to give a polymer solution of about 22% solids. This material functions as a conductivity aid for electroconductive paper when formulated in the manner described in Example 2(c). It can be given a dielectric coating layer as described in Example 2(d).

Ammonium Sulfonated Polystyrene plus Ammonium Sulfamate

A definite need exists for a water-soluble conductive material for treating sheet material used in electroconductive and photoconductive reproducing, copying and printing processes, which material will be compatible with the other papermaking additives and will maintain a high electroconductivity under widely varying conditions of temperature and humidity.

This is achieved in the electroconductive sheet material described by *D.R. Cahill; U.S. Patent 3,486,936; December 30, 1969; assigned to Monsanto Company* It comprises a fiber base having dispersed therein from 0.5 to 15.0 pounds per 3,000 square feet of fiber base of a water-soluble conductive material comprising from 25 to 100 (preferably 50 to 90) weight percent of sulfonated polystyrene and salts thereof and from 0 to 75 (preferably 10 to 50) weight percent of salt of sulfuric acid and its derivatives (preferably sulfamic acid) where in each case the weight percent is based on the total weight of water-soluble conductive material and where the cation in the respective salts is selected from the group consisting of ammonium, alkali metals and amines.

The term ammonium is meant to include substituted ammonium or amines such as ethanolamine, morpholine, picoline, etc. which are generally recognized as derivatives or substituted ammonium.

The conductivity tests described are carried out according to the procedure of D.G. Brubaker, "Testing of Electrophotographic Papers and Coating Formulations", *TAPPI*, Vol. 46, No. 5, May 1963, pages 312-316. Unless otherwise indicated any treated or untreated paper stock used is conditioned for five days at constant temperature and humidity prior to testing for electroconductivity. For the sake of uniformity and ease of comparison unless otherwise indicated the paper stock used is Riegel EC35XE base stock.

Paper treated with a solution of an ammonium salt of sulfonated polystyrene or with ammonium sulfamate shows improved conductivity over untreated paper. However, it has been discovered that a mixture of sulfonated polystyrene and the salts thereof and the salts of sulfuric acid and its derivatives may be used to great advantage as the conductive material in photoelectric or electroconductive paper. These mixtures show electroconductivity that is better than either material used alone at any given weight. Furthermore, the amount of the inorganic salt used to achieve a desired level of conductivity is well below those amounts which attract excessive moisture and give a limpness or raggy feeling to the sheet. Preferably one would use sulfonated polystyrene having a molecular weight in the 50 to 600,000 range and at least 1.0 unit of sulfonic acid or salts thereof per repeating styrene unit. The polymeric material should have a solubility of at least 2% by weight in aqueous solutions at 25°C.

The water-soluble electroconductive materials may be introduced into the sheet material during or after the manufacture of the sheet material. In the former instance the materials are added to the pulp or the nearly dried paper after its formation on a Fourdrinier machine. After the formation of the paper, the materials are applied to the sheet by spraying, knife coating, roller coating, transfer rolls, etc. Also, the sheet may be brought into contact with an aqueous or methanol solution of the electroconductive materials. When the paper is to be coated with aqueous based film forming materials such as the emulsion of Examples 1 and 2, the material may be incorporated into the aqueous composition.

Example 1: The paper stock described was treated with the following slurry to give an applied weight of 4 pounds per 1,000 square feet.

	Pounds
Ammonium salt of sulfonated polystyrene	2.0
Gelva TS-30 polyvinyl acetate emulsion (dry basis)	1.0
Clay (weight of dry solids)	1.0
Water	16.0

The ingredients were readily formulated and no incompatibility was observed in the slurry. Paper stock treated with the above slurry gave the following conductivities: 44% RH, 4.8 megohms/square; and 12% RH, 90 megohms/square.

Example 2: Example 1 was repeated here except that no clay was used. A dry coating weight of 2.0 lb/3,000 ft² gave the following conductivities: 44% RH, 6.0 megohms/square; and 12% RH, 870 megohms/square. The compatibility of the mixture was excellent. Furthermore, the polyvinyl acetate emulsion used served as a coating on the paper which enhanced the solvent hold-out of the surface so treated. This latter feature provides a more uniform surface upon which to coat a solvent based electroconductive film which allows better copies to be made.

Example 3: This example illustrates the use of a 65/45 weight ratio of sodium salt of sulfonated polystyrene/sodium sulfate as a water-soluble electroconductive material for treating paper stock. Paper stock treated with a dry solids weight of 2 lb/3,000 ft² give the following conductivities: 44% RH, 4.5 megohms/square; and 12% RH, 700 megohms/square.

Quaternary Ammonium Derivatives of 2-Chloromethyl-1,3-Butadiene Polymers

G.D. Jones, G.R. Geyer and M.J. Hatch; U.S. Patent 3,544,532; December 1, 1970; assigned to The Dow Chemical Company describe cationic polymers containing a plurality of moieties of the general formula:

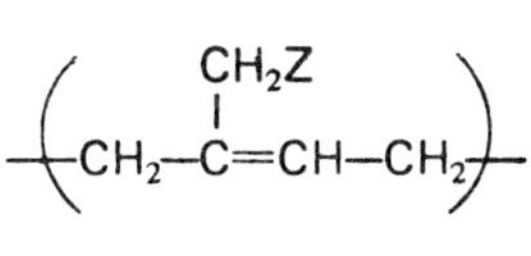

where Z is an amino, ammonium or sulfonium group. These are obtained by polymerization of 2-methylene-1,3-butadiene amino, ammonium or sulfonium monomers or by reaction of poly(2-chloromethyl-1,3-butadiene) with an appropriate reagent to substitute the cationic group Z for the reactive chloro group. The resulting polymers are useful as flocculants, dispersants and thickeners, and wet-strength additives for paper. The quaternary ammonium polymers are effective electroconductive additives for paper.

Example 1: (a) Quaternary Ammonium Monomers — 2-Chloromethyl-1,3-butadiene was obtained by fractional distillation of chlorinated isoprene. A mixture, BP 76° to 78°C/300 mm, contained by vapor phase chromatographic analysis 80% 2-chloromethyl-1,3-butadiene with 20% cis- and trans-1-chloro-2-methyl-1,3-butadiene and isoprene hydrochlorides. By further fractionation, 2-chloro-1,3-butadiene with a minimum purity of about 95% was obtained.

To a solution of 29.5 parts (0.29 mol) of 96.7% 2-chloro-1,3-butadiene in 80 parts of absolute ethanol was added at room temperature 70 parts (1.2 mols) of trimethylamine in 160 parts of ethanol. After reacting for 2 hours at room temperature, the solvent and excess amine were stripped from the product in vacuo to give 37.3 parts (80% yield) of a hygroscopic white crystalline solid which readily dissolved in water. The infrared spectrum and elemental analysis of the product were consistent with the structure:

$$\begin{array}{c} \qquad CH_2N^+(CH_3)_3Cl^- \\ | \\ CH_2{=}C{-}CH{=}CH_2 \end{array}$$

that is trimethyl(2-methylene-3-butenyl)ammonium chloride. The ultraviolet spectrum of a dilute aqueous alcohol solution showed a peak of 220 mμ characteristic of a 1,3-diene with an extinction coefficient of about 28,000 cm^2/mol. From analysis of the nuclear magnetic resonance (NMR) spectrum in D_2O solution, the monomer contained less than 5% dimer.

(b) Quaternary Ammonium Polymers — (a) A mixture of 31.4 parts of the quaternary trimethylammonium monomer described in Example 1(a), 7.5 parts of water, and 3 parts of a 5% solution of disodium ethylenediaminetetraacetate was adjusted to pH 5 and deaerated. Then 0.74 part of a 10% aqueous solution of tert-butyl hydroperoxide was added and the mixture heated for 16 hours at 60°C. The resulting trimethylammonium polymer (TMA-2) had a viscosity of 1.30 cp as a 5% solution in dilute (4%) sodium chloride at pH 5 and 25°C. No monomer was evident in the NMR spectrum.

Example 2: (a) Poly(2-Chloromethyl-1,3-Butadiene) Intermediate — An aqueous emulsion was prepared by vigorously agitating a mixture of 260 parts of 80% 2-chloromethyl-1,3-butadiene, 0.12 part of tert-dodecyl mercaptan, and an aqueous solution of 1.4 parts of sodium lauryl sulfate, 0.3 part of potassium persulfate and 0.3 part of sodium bicarbonate in 140 parts of water. Polymerization was achieved by heating the emulsion at about 50°C for 40 hours. Then the polymer was precipitated by pouring the emulsion into an excess of methanol with vigorous agitation. The finely divided, white product was recovered by filtration, washed with methanol and then air-dried.

The poly(2-chloromethyl-1,3-butadiene) was insoluble in water and methanol, but dissolved in methylene chloride, toluene and tetrahydrofuran. It contained 30.3 weight percent side-chain chlorine as determined by reaction with excess trimethylamine and 31.5 weight percent total chlorine by elemental analysis. The higher total chlorine content indicates copolymerization of some 1-chloro-2-methyl-1,3-butadiene. The polymer had an average molecular weight of 35,000 as determined by the membrane osmotic pressure of a tetrahydrofuran solution. Spectral studies indicate that the polymer has predominately the 1,4-addition structure.

(b) Ammonium Derivatives — (a) To about 250 parts of an isolated poly(2-chloromethyl-1,3-butadiene) containing 30.3 weight percent side-chain chlorine was added 720 parts of 25 weight percent aqueous trimethylamine (30% excess) and the mixture was agitated at room temperature until the polymer completely dissolved. The resulting quaternary ammonium derivative was precipitated by pouring the aqueous solution into an excess of acetone with vigorous stirring. A sample of the hygroscopic cationic polymer, purified by reprecipitation from aqueous solution contained 19.2 weight percent ionic chloride indicating a purity of about 90% as the quaternary trimethylammonium derivative (TMA-6).

Example 3: Use as Electroconductive Additives — The quaternary ammonium polymers described in Examples 1 and 2 are effective electroconductive additives for paper. In tests made according to the general method described in ASTM D257-61, standard test sheets of 51 pounds bleached sulfite paper were dipped in an aqueous solution of the quaternary ammonium polymers until a 5 to 15 weight percent polymer pickup was obtained. The treated sheets were dried for 5 minutes at about 110°C weighed, and cut into standard test strips. The strips were conditioned at room temperature and the desired relative humidity (70 RH) for 24 hours. Then the electroconductivity was measured using strips treated with a commercial quaternary ammonium electroconductive resin as a control standard.

Typical results given in the table compare the effectiveness of the polymerized trimethyl(2-methylene-3-butenyl)ammonium chloride (TMA-2) and the trimethylamine derivative of poly(2-chloromethyl-1,3-butadiene) (TMA-6) with the commercial additive in both surface and volume resistivity.

Electroconductive Additives

A. Surface Resistivity

Additive	Weight Percent	Surface Resistivity, ohms 7-10% RH	30-33% RH	56-60% RH
Commercial	7.0	2.6×10^{9}	3.5×10^{8}	1.8×10^{7}
TMA-2	10.0	4.4×10^{11}	2.8×10^{8}	1.1×10^{7}
TMA-6	8.0	-	3.4×10^{10}	-

B. Volume Resistivity

Additive	Weight Percent	Volume Resistivity, ohms 7-10% RH	30-33% RH	60% RH
Commercial	7.0	1.2×10^{13}	8.1×10^{11}	5.7×10^{9}
TMA-2	8.2	1.9×10^{13}	4.0×10^{11}	2.9×10^{9}
TMA-6	16	-	4.8×10^{9}	1.8×10^{7}

Chrysolite Asbestos, LiCl, and Polyelectrolyte Complex Resin

The value of using a colloidal chrysolite asbestos to impart conductivity to cellulosic paper has been recognized. U.S. Patent 3,337,392 discloses the use of such an asbestos and polyvalent metal salts in the formation of a conductive paper matrix. However, it has been discovered that the optimum advantage in utilizing colloidal chrysolite asbestos is to be obtained in avoiding the dispersal of the material throughout the paper web, i.e., in restricting the presence of the asbestos to the surface of the paper.

A distribution of the conductivity-imparting material in the substrate stock is undesirable for a number of reasons. For example, this mode of employing the conductivity-imparting material increases the likelihood of strike-through of the material which interferes with image formation on the underlying sheets in a pad of sheets. Also, confining a given quantity of conductivity-imparting material to the surface of a sheet of substrate paper to yield high surface conductivity results in better image formation than when the same quantity of material is dispersed through the bulk of the sheet which dispersal, while elevating the bulk conductivity, reduces the surface conductivity.

R.A. Cross and J.F. Hegarty; U.S. Patent 3,607,377; September 21, 1971; assigned to Amicon Corporation provide means for achieving a coating composition comprising colloidal chrysolite asbestos which coating can be retained on the surface of a paper, thereby contributing to avoiding the aforementioned problem, and to obtaining of a highly satisfactory surface-conductive paper for use as a substrate for papers used in electrostatic-printing processes over a wide range of humidity levels.

The colloidal chrysolite asbestos preferred for use is that known as Coalinga asbestos, obtained from a deposit located near Coalinga, California. Asbestos mined at this location is short-fibered chrysolite asbestos and would be classified between grade 5 and grade 7 according to the Canadian Standards Classification. Coalinga asbestos, however, has a much more uniform distribution of individual fiber lengths and diameters than do comparable grades of Canadian chrysolite asbestos. In addition, the surface area of Coalinga asbestos is substantially greater than that of the Canadian type of asbestos.

It is surprising that a coating of Coalinga asbestos provides really good electrostatic-printing paper over a rather wide range of humidities without the addition of salt. However, to achieve the broader advantages of the process, some salt and/or polyelectrolyte complex resin is advantageously included in the coating formulation.

The most advantageous salt is the monovalent lithium chloride not only because it provides excellent low-humidity moisture retention properties, but also because it is retained in the asbestos-gel coating to a surprising extent.

The polyelectrolyte complex resins useful in the process include those such as described in an article entitled "Polyelectrolyte Complexes" by Alan S. Michaels in the October 1965 issue of *Industrial and Engineering Chemistry,* a publication of the American Chemical Society. The polyelectrolyte complex resin is formed by the ionic association of a sulfonate polyanion such as poly(sodium styrene sulfonate) and a quaternary ammonium-type polycation such as poly(vinylbenzyltrimethyl ammonium chloride).

A coating composition, suitable for use in imparting conductivity to paper at low humidity levels, consists essentially of: 10 parts of colloidal chrysolite asbestos, 2 to 10 parts of lithium chloride, 0 to 5 parts of a polyelectrolyte complex resin (preferably at least 2), and a quantity of water in which to disperse the foregoing ingredients.

When the coating comprises a colloidal chrysolite asbestos in the quantity of from 0.5 to 5 pounds per ream of paper sheet, the resultant sheet has a surface resistivity of less than 10^{10} ohms at a relative humidity of 15% and a volume resistivity of more than about 10^{10} ohm-cm at 15% relative humidity.

Example (a) A chrysolite asbestos of the Coalinga type was dispersed in water to form a slurry containing 5% of the asbestos. The pH of the slurry was lowered to about 4.3 with a small quantity of acetic acid. Thereupon the pH of the slurry was raised to 5.5 by adding a small quantity of barium hydroxide thereto. Then a single drop of a 31% aqueous solution of poly(vinylbenzyltrimethyl ammonium chloride) was added to the slurry.

Separately, a dispersion of 325 mesh anionic polyelectrolyte complex resin Ioplex, was also dispersed to form a 5% solids suspension in water. These two dispersions, i.e., the asbestos and resin dispersions, were mixed in quantities so that the resultant dispersion contained five parts asbestos and one part polyelectrolyte complex resin. Thereupon enough lithium chloride was dissolved in the slurry so that the resin-to-salt ratio was 1 to 1.

The slurry was drawn down onto a sheet of Crocker Videograph paper using a number 18 wire-wound casting rod of the type well known in the paper-coating art. The resulting coating weight, on a dry basis, was about 1.7 pounds per ream of paper. This coated paper was dried for 5 minutes in a circulating air oven at 80°C. A smooth, colorless, coating was observed on the dried sheet.

Conductivity of the paper was tested by placing the sheet in a dry box having a relative humidity of 16%. After conditioning the paper for about 16 hours in this dry environment, its surface and volume resistivity were tested with a Keithley 6105 Resistivity Adapter and a Keithley 621 Electrometer using a Sorenson High Voltage Power Supply, respectively. Surface resistivity was 2×10^9 ohms/sq and volume resistivity was 10^{11} ohm-cm. These values indicated that there was little or no strike-through of the coating composition below the surface of the paper.

(b) The procedure of (a) was repeated excepting that the polyelectrolyte complex resin was omitted from the papercoating composition. The surface resistivity of the resulting paper was 5×10^9 ohms/sq.

(c) The procedure of (a) was again repeated excepting that (1) both the polyelectrolyte complex resin and the lithium chloride salt were omitted from the formulation; (2) the coating contained 2.3 lb per ream of asbestos; and (3) the testing was carried out at 45% relative humidity. Surface resistivity was 1×10^9.

ELECTROPHOTOGRAPHIC PAPERS

Binder, Clay, Sulfate Salt Surface Size for Lightweight Bulked Papers

Lightweight bulked papers containing aminoplast thermosetting resins, thermoplastic gaseous filled resinous spheres, or natural polymeric substances as bulking agents are unsuitable for electrophotography, because the bulking substance appears at the surface of the paper.

C.J. Green, Jr., G. Treier and R.H. MacClaren; U.S. Patent 3,884,685; May 20, 1975; assigned to Xerox Corporation provide a lightweight surface sized electrophotographic copy paper having excellent curl and friction characteristics for use in electrophotographic machines. This is achieved by incorporating into the paper a low density bulking material, and subsequent surface sizing of the lightweight paper. Electrophotographic copy papers are readily prepared by the addition of up to about 10% of the low density bulking material, based upon the weight of the dry pulp, to the pulp slurry. The resulting lightweight paper is then surface sized with a composition comprising a binder, a salt, preferably in amounts from about 1 to 50% by weight based upon the amount of binder, and 40 to 50% clay based on the total coating composition, but such that the combination of salt and clay does not exceed 60% based on the total coating composition.

Any commercially available binder products conventionally used in surface sizing compositions may be used. Typical binders include starch, starch derivatives, polyvinyl alcohol, polystyrene, and mixtures thereof. Because of its antiscorch properties, polystyrene latex in a weight ratio of from about 40% or more is a particularly useful binder in the sizing. Preferable salts include the sulfate salts of any metal. Typical salts which give exceptional results are magnesium sulfate, zinc sulfate and sodium sulfate and the hydrates thereof. Any type of coating clay may be used.

For optimum properties for use in electrophotographic machines the paper should comprise over 90% cellulose fiber to avoid roughness in the paper surface. Therefore, the lightweight paper preferably should contain from about 0.5 up to 5% bulking material.

The particular material used for bulking the paper is not critical and may be synthetic resins of the type described in U.S. Patents 3,293,114; 3,210,239; 3,128,478; 2,929,106; and 2,797,201. Natural bulking substances, such as foamed starch and cellulosic materials, can be used with similar results. In addition, inorganic bulking substances such as hollow glass, ceramic and metal microspheres can be used.

In the preparation of the lightweight electrophotographic copy paper, it is found that bulking material accumulates on the surface of one side of the paper. In this condition the coefficient of friction of that particular side of the paper is very high. Therefore, while not a requirement in the preparation of the electrophotographic copy paper, it is preferable to subject the side of the paper having the bulking material on its surface to a mechanical rubbing treatment with a cotton fabric material before the step of surface sizing whereby flattening, abrading, or breaking of the excess bulking material occurs at the surface of the paper. By means of the mechanical treatment of the paper those loose or protruding particles of bulking material are removed or broken thereby rendering the subsequent surface size more effective.

Example 1: An electrophotographic copy paper is prepared as follows. A surface sizing is prepared by adding 100 parts by weight of Penford Gum 280, an ethylated starch; 50 parts by weight Vinol 125, a polyvinyl alcohol; and 18 parts sodium sulfate to 1,500 parts of water. A 220 parts by weight predispersed clay solution, made with Nuclay, which was prepared by adding enough clay to water to obtain a 60% solids mixture, is then added to the coating composition.

A slurry is prepared by adding 1,000 pounds of dry softwood kraft fiber and 500 pounds of hardwood kraft fiber to 50,000 gallons of water in a hydropulper. The fiber is refined to a 350 Canadian Standard Freeness in a Jordan. Thereafter 90 pounds of Experimental Resin CX 7058-1, which are gaseous resinous spheres, are added to the slurry. The slurry is diluted to 0.5 to 1% by weight of fiber and then fed into the head box of a Fourdrinier paper machine. The resulting paper passes through the dryers and at the size press the above prepared sizing composition is applied to the paper. The sized sheet is then dried and very lightly calendered.

The resulting surface sized lightweight electrophotographic copy paper is cut to fit a Xerox 720 machine. A stack of 200 sheets is then placed in a Xerox 720 machine and set to copy a standard pattern 200 times. There is no jamming of the machine and all 200 copies are of excellent quality demonstrating no discoloration.

Example 2: A urea-formaldehyde foam (0.8 lb/ft^3, 150 micron cell diameter) is pulverized in an Abbe mill with a one-sixteenth inch screen plate. Discrete amounts of the disintegrated foam are then blended with a slurry of 1,000 lb of dry softwood kraft fiber, 500 lb of hardwood kraft fiber, and 5,000 gallons of water, the slurry being prepared in the same manner as Example 1. A total of about 2% by weight of the disintegrated thermosetting resin material is added to the slurry. The slurry is then diluted to about 1% by weight of fiber and a sized sheet prepared in the same manner as Example 1 using the same sizing composition. The resulting lightweight paper is tested in a Xerox 720 machine in the same manner as Example 1 with the same advantageous results.

Inorganic Salt/Polyhydric Alcohol-Based Coatings

The electrophotographic paper described by *J.K.J. Cheng; U.S. Patent 3,615,403; October 26, 1971; assigned to GAF Corporation* is suitable for use in both wet and dry process machines. It was found that the replacement of the conductive resin in the conductive-treating solutions by an inorganic salt-polyhydric alcohol system results in coated papers having greatly improved properties. Among these improvements are: excellent solvent and dispersant hold-out, good conductivity, acceptable moisture retention and minimal sensitivity to humidity changes over a relative humidity range of 16 to 80%.

These results are all the more surprising since previous evaluations using inorganic salts alone had not only been unsuccessful but were, in fact, shown to have a deleterious effect on the coated paper due to the resulting sensitivity to humidity. Thus, totally unexpected results arise from the unique combination of a polyhydric alcohol having moisture-regulating properties with an inorganic salt.

A typical conductive barrier film-forming solution is composed of the following ingredients in the range and preferred quantities listed on the following page.

Ingredient	Range, parts	Preferred, parts
Inorganic salt	2-4	2.25-3
Polyhydric alcohol	2-4	2-4
A polyvinylpyrrolidone polymer or a polyvinylpyrrolidone-vinyl acetate copolymer	0.25-5	3.3-4.3
Polyvinyl acetate emulsion	2.5-5	3.5-4.5
Acetylated starch	2-6.5	2.2-3
Polyvinyl alcohol	2-6.5	3-5
Water to make	100	100

The inorganic salt is preferably one containing a monovalent cation. One example of such compound is sodium chloride of 99.95% purity, Culinox 999.

Suitable polyhydric alcohols include glycerol, glycols such as ethylene or propylene glycol, 1,4-butanediol and polyethylene glycols. Since the paper surface in liquid toner systems must be completely immersed in the developer solutions, a barrier layer coating must be applied thereon. It has been found that an essential constituent in the coating compositions is a polyvinylpyrrolidone polymer. Examples of some vinylpyrrolidone polymers which are acceptable include: poly-N-vinylpyrrolidone, N-vinylpyrrolidone-vinyl acetate copolymer (10:90), N-vinylpyrrolidone-vinyl acetate copolymer (25:75), N-vinylpyrrolidone-vinyl acetate copolymer (30:70), N-vinylpyrrolidone-ethyl acrylate copolymer (30:70), and N-vinylpyrrolidone-ethyl acrylate copolymer (30:70) mixed with poly-N-vinylpyrrolidone-vinyl acetate copolymer (25:75). The ratios given in the parentheses indicate the proportions by weight of the component monomers.

Additional ingredients, polyvinyl acetate, acetylated starch and polyvinyl alcohol add body and gloss properties to the paper substitute or impart the required viscosity to the coating compositions.

Example 1: Preparation of Coating Solutions — The coating compositions were formulated by standard chemical procedures, i.e., the salt was dissolved in water and added to a starch-polyvinyl alcohol slurry. This mixture was then heated to and maintained at 195°F for 15 minutes. The resulting solution was cooled to 120°F and the other ingredients were added followed by dilution of the mixture to the desired percent solids (15 to 35%). The formulations are identified by letter for purposes of convenience of reference to indicate these compositions in the examples below.

	Formulation			
	A	B	C	D
Mol ratio salt:glycol	2.24	2.24	1.3	0.81
NaCl	3	3	-	-
NH_4Cl	-	-	2.25	-
$NaNO_3$	-	-	-	2.25
Polyvinylpyrrolidone	0.5	0.5	-	-
Polyvinylpyrrolidone/vinyl acetate	-	-	3.3	3.3
Polyvinyl acetate	5.5	5.5	3.45	3.45
Acetylated starch	3	3	2.2	2.2
Polyvinyl alcohol	3	3	3.75	3.75
Glycerol	2.1	2.1	3.0	3.0
H_2O to make	100	100	100	100

Example 2: (a) Solvent and Toner Holdout Test — An empirical test was developed to measure the efficiency of the various coatings in resisting both solvent

and liquid dye penetrations. A frame assembly, having an 8½ x 11 inch sheet of coated paper tightly clamped thereon, is positioned on a support with the surface of the paper at a 45° angle to the horizon. A funnel connected through a valve to a hypodermic syringe whose tip delivers a 2 mm diameter drop is positioned above the paper with the tip 12 mm from the paper's surface and perpendicular to the horizon.

In operation, a drop from the pipette strikes the paper surface and travels until completely absorbed or dried. The length of this free travel path is directly relatable to the efficiency of the various coatings. Measurements are made on both the wire and felt sides of the paper.

(b) Resistivity Test — Data were also obtained on both surface and bulk resistivity. Surface resistivity was calculated from measurements made of the voltage developed between a pair of one inch brass electrodes spaced one inch apart when placed on a paper sample mounted on an insulated surface. Bulk resistivity, i.e., the resistivity through the thickness of the paper, was calculated in an analogous manner with the paper placed on a conductive surface.

In this procedure, however, the recorded voltage was that developed between the same pair of electrodes so positioned that one was on the paper and the other was on the conductive surface. These voltage measurements which were, in turn, converted into resistivity, were determined over a range of humidities. Since they are indicative of the static susceptibility of the paper surface, they are, of course indicative of moisture content and, hence, humidity sensitivity.

Example 3: Evaluation — The beneficial effects obtained in using coating compositions containing various inorganic salts in combination with a polyhydric alcohol(glycerol) are shown. Coating compositions prepared as described in Example 1 were applied to various paper surfaces using the air knife techniques, at such a rate as to yield a coating, when dry, of 4.5 pounds for 3,000 ft^2.

Solvent and toner hold-out were determined by the procedure of Example 2. The coatings offered excellent resistance to both solvent and toner absorption or penetration. Surface and bulk resistivities, determined at several humidities using the procedure described earlier, indicated that the coated papers were acceptable for electrophotographic use in either wet (liquid) or dry development processes. Details as to the compositions of the coatings, the surface coated and the test data obtained are tabulated below.

							Resistivity (ohms x 10^9/in^2)								
			Holdout (mm)				20% RH			50% RH			80% RH		
Formulation	Salt	Paper Surface	Solvent		Toner		Surface			Surface			Surface		
			Felt	Wire	Felt	Wire	Felt	Wire	Bulk	Felt	Wire	Bulk	Felt	Wire	Bulk
A	NaCl	Diazo	65	70	66	71	-	-	-	0.1	0.1	0.05	-	-	-
B	NaCl	Pattern	62	57	60	58	-	-	-	0.13	0.15	0.047	-	-	-
C	NH_4Cl	Diazo	76	73	78	72	11	8	1.7	0.17	0.11	0.058	0.0012	0.0012	0.00089
D	$NaNO_3$	Diazo	82	79	73	74	20	2	2.7	0.36	0.37	0.10	0.0019	0.0018	0.00089

PHOTOGRAPHIC PAPERS

TUB SIZING WITH CROSSLINKED POLYMER LATEX

W.H. Griggs, M.K. Eldridge and D.R. Fulton; U.S. Patent 3,582,337; June 1, 1971; assigned to Eastman Kodak Company have developed a method for tub sizing photographic paper base which imparts unexpected properties to the paper in addition to improved stiffness.

The process involves tub sizing this paper with an aqueous latex of an emulsion polymerized mixture of vinyl acetate and a crosslinking monomer such as N-methylol acrylamide or hydroxyethyl acrylate. The crosslinking monomer is, thus, a polymerizable vinyl compound containing the vinyl group for copolymerizing with the vinyl acetate.

In addition, the monomer contains a lower hydroxyalkyl group, such as a hydroxyethyl or methylol, capable of reaction with similar groups on other copolymer chains in the reaction mixture to produce a crosslinked polymer latex. As a result, papers of improved stiffness are obtained and sensitive coatings applied thereto are free of mottle even when lightweight paper stocks have been used. Resistance to solution penetration is also good. The sized paper surface has no adverse effect on the light-sensitive layers.

Crosslinking vinyl monomers containing a lower hydroxyalkyl group especially useful for emulsion polymerization with vinyl acetate are: N-methylol acrylamide, N-methylol methacrylamide, glycidyl acrylate, glycidyl methacrylate, hydroxyethyl acrylate, hydroxyethyl methacrylate, hydroxypropyl acrylate or hydroxypropyl methacrylate.

The crosslinking monomer is used particularly in an amount necessary to obtain the desired hardness in the vinyl acetate polymer and stiffness in the paper. About 0.5 to 10% by weight of the crosslinking monomer is useful with about 99.5 to 90% by weight of vinyl acetate.

N-methylol acrylamide and N-methylol methacrylamide are especially useful for emulsion polymerization with vinyl acetate in these proportions. The emulsion polymerization is carried out in water in a well-known manner by heating in the presence of a surfactant and catalyst so as to produce a latex of about 40 to 50% solids by weight.

The surfactant and catalyst are selected so as to be photographically acceptable i.e., inert to light-sensitive photographic layers. Anionic surfactants such as sodium lauryl sulfonate are useful. Water-soluble catalyst, e.g., potassium persulfate, can be used with heating to about 70°C or sodium meta-bisulfate with potassium persulfate and heating to about 40°C. The resulting latex should then be diluted to about 30 to 35% solids for use in tub sizing the paper.

The paper is tub sized with the crosslinked vinyl acetate polymer latex in the usual manner on the paper machine. The latex of about 30 to 35% solids concentration at about 90°F is applied freely to both sides of the paper by passing the paper through a bath containing the latex or it can be sprayed onto the paper. Thereafter, excess size solution can be removed by passing the paper through press rolls and drying and calendering the paper as usual.

The paper can be expected to contain from about 5 to 25% crosslinked polymer depending upon the solids concentration of the latex used, speed and temperature of applying the latex. The paper stock to which the vinyl acetate polymerization latex is applied can vary widely as to fiber content but a sheet of good formation is desirable. The weight can vary, for example, from about 4 to 50 pounds per 1,000 square feet.

The paper sized with the crosslinked vinyl acetate polymer is suitable for sensitizing directly with photographic emulsions. However, if desired, the usual baryta coating may first be applied. Also, the sized paper can be coated on each surface with hydrophobic polymer layers before coating photographic emulsion layers.

For this purpose, polyolefins, especially polyethylene and polypropylene, are advantageously extruded onto the paper from a hot melt. The adhesion of the polymer layers applied to the sized paper at high speeds can be increased by activating the paper surface with corona discharge.

Photographic emulsions, such as gelatin-silver halide emulsions are advantageously adhered to the surface of the polyolefin coated paper by activating the polyolefin surface with corona discharge. The photographic emulsions which are coated upon the sized paper surface are the conventional black-and-white emulsions or a plurality of known differently sensitized emulsions for color photography.

Example 1: A 37 pound softwood bleached sulfite aluminum stearate internally sized paper stock is tub sized at about 90°F with a 35% solids latex of a crosslinked vinyl acetate polymer resulting from emulsion polymerizing about 5 parts of N-methylol acrylamide and 95 parts of vinyl acetate in the presence of catalyst and surfactant in water.

The same paper stock is tub sized with 4% starch solution and with water alone for comparison. The stiffness and acid penetration values of the resulting papers are shown in the table on the following page.

Size	Thickness (mils)	Taber Stiffness (length)	Valley Acid Penetration (seconds)
Water	8.2	19.2	18
Starch	8.3	20.5	23
Polymer	8.0	26.0	832

The improvement in stiffness and penetration is apparent. The polymer sized paper coats uniformly with gelatin-silver halide photographic emulsion, handles well in processing and no mottle is observed in the emulsion layer whereas the water and starch sized papers exhibit significant mottle in the emulsion layer.

Example 2: 27 pound paper not internally sized and not calendered is tub sized at about 300 fpm with a 30% solids latex of the vinyl acetate polymer of Example 1 at 85° to 90°F and for comparison tub sized with water and 4% starch solution. The papers have the following properties.

Size	Taber Stiffness L	C	Thickness (mils)	Valley Acid Penetration (seconds)
Water	6.6	4.0	5.8	91
Starch	7.4	-	5.8	197
Polymer	12.0	6.0	5.8	607

The polymer sized paper is extrusion coated with about 8 pounds polyethylene on each surface and sensitized with a black-and-white gelatin-silver halide emulsion without production of mottle. The resulting polymer paper has a stiffness comparable to a 40 pound paper which has been polyethylene coated and sensitized under the same conditions.

Example 3: A paper is prepared as in Example 2 except the initial paper stock is internally sized with aluminum stearate. The stiffness of the paper before polyethylene coating L/C is 12.1/6.4 to 13.5/6.5 compared to stiffness 8.0/5.0 for the same paper not polymer sized.

When this polyethylene coated paper is sensitized and processed, it shows improved resistance to edge penetration by the processing chemicals. The internally sized paper, tub sized with the crosslinked polymer latex and polyethylene coated, is especially useful for coating with differently sensitized emulsions for color photography. The emulsions and polymer layers of the element adhere well to the sized paper and the paper has good feel and handling properties.

Example 4: This example illustrates an important aspect of the process where lightweight photographic paper stocks of about 6 to 10 pounds are sized with the vinyl acetate polymer latex to provide photographic paper bases which can be sensitized without producing mottle. If the latex sizing is not used, mottle-free paper is usually not obtainable.

A thin, stiff black-and-white emulsion sensitized photographic paper is required in a certain process for direct recording of an oscillograph tracing. A common 10 pound paper stock which had been emulsion coated could not be passed through the apparatus successsfully. The 10 pound paper is tub sized as described in Example 2 with the 30% polymer latex to obtain a paper having the following properties shown in the table on the following page.

Taber Stiffness

	L	C
Unsized	3.4	2.1
Sized	4.9	3.1

The paper sensitizes uniformly with the gelatin-silver halide emulsion and handles satisfactorily in the apparatus when recording the oscillograph tracing. Mottle is not evident in the processed emulsion layer. In the above examples the acid penetration values are obtained by a well-known method where the instrument is calibrated to 40 milliamps and the test is run to a 20 milliamp end-point. The penetrating solution is an aqueous solution of acetic acid, water and sodium chloride.

In the above examples the crosslinked vinyl acetate polymer sized paper can be coated with gelatin baryta layers prior to sensitizing with photographic emulsions. In this case it is advantageous to include a small amount of the vinyl acetate polymer latex into the aqueous gelatin baryta dispersion. About 10 to 15% of the polymer based on the weight of gelatin present (dry weight basis) is useful with the result that adhesion of a gelatin-silver halide emulsion layer to the baryta coating is improved.

TRANSPARENT PAPER USEFUL AS UNDERLAY FOR PHOTOSENSITIVE COATING

Transparent paper has been known for a long time and is particularly used for drawing paper, lampshade material, and photoreproduction papers for the production of, for example, diazo copies or the like.

G. Budde; U.S. Patent 3,497,370; February 24, 1970; assigned to Felix Schoeller, Jr. describes a method for making transparent paper which comprises treating the paper with a composition which comprises an emulsion or a melt of about 80 to 95% by weight of paraffin oil, about 5 to 20% by weight, preferably about 10% by weight of stearic acid, and about 50 to 300%, preferably 200%, by weight of resin alcohol such as abietyl alcohol.

By use of resin alcohol in combination with the paraffin oil and the stearic acid there is avoided a greasing-off or giving off of transparentizing agents to photographic baths and to papers which come in contact with the transparent paper. The percentage of resin alcohol is in reference to the total amount of paraffin oil and stearic acid.

The transparentizing agent can be used in the form of a solution of esters and organic hydrocarbons. Also, a melt through which the paper is drawn can be used. It can also be added to the paper as an aqueous emulsion inside a paper machine. The paper to be made transparent can be of cellulosic fibers or of synthetic fibers or of a mixture of both types of fibers.

The papers produced according to the process are stable to light, indifferent with respect to photosensitive coatings and stable with respect to photographic baths or chemicals such as are used in diazotyping. There is no reduction of the transparency even with relatively long storage and in processing in photographic baths. The papers produced do not stick together and are practically odorless.

Example 1: Use of Solution — (a) A roll of photo crude paper of 55 grams per square meter is drawn at a rate of 20 meters per minute through a solution of 20 parts of paraffin oil and 2 parts of stearic acid in a mixture of 60 parts of acetic ester and 13 parts of toluene; and the excess of impregnating agent is stripped off with the aid of scrapers.

After the evaporation of the solvent, the paper is immediately suited, without application of additional layers, for coating with photographic silver halide emulsions. The optical density of the transparent paper as measured with an opacimeter amounts to 0.28. The opacity of the crude paper before the transparentizing was 0.50.

(b) A roll of crude paper, 80 grams per square meter is drawn at a rate of 15 meters per minute through a solution consisting of: paraffin oil, 12.7% (90.7% based upon total weight of paraffin oil and stearic acid); stearic acid, 1.3% (9.3% based upon the total weight of paraffin oil and stearic acid); resin alcohol (abietyl alcohol), 26.0% (185% based upon the total weight of paraffin oil and stearic acid); acetic ester, 45% and; toluene 15%.

The excess of impregnating agent is stripped off with the aid of scraper blades. After the evaporation of the solvent the impregnated paper stood for a few days in order to give the impregnating agent situated on the surface of the paper an opportunity to diffuse into the hollow spaces in the paper formerly occupied by the solvent. The optical density of the transparent paper is 0.40 and that of the crude paper is 0.64.

Example 2: Use of Aqueous Emulsion in Papermachine — A photographic crude paper of 55 grams per square meter is drawn, after the sheet formation on the papermachine, with a dry content of about 60%, through an emulsion consisting of 20 parts of paraffin oil, 2 parts of stearic acid, 2 parts of Emulsogen A (a fatty acid polyglycol ester recommended for emulsifying mineral oils and fatty oils) and 76 parts of water. The excess impregnating agent is squeezed out by rubber rollers. The paper is then dried within the papermachine to a moisture content of about 6% of water. The optical density of the impregnated paper amounts to 0.32.

Example 3: Use of Melt — A photographic crude paper of 55 g/m^2 is drawn through a melt which consists of 70% (233% based upon the total weight of paraffin oil and stearic acid) of resin alcohol, 27% (90% based upon the total weight of paraffin oil and stearic acid) of paraffin oil and 3% (10% based upon the total weight of paraffin oil and stearic acid) of stearic acid. The excess of impregnating agent is stripped off with the aid of scrapers heated to 60°C. The optical density of the transparent paper is 0.18 and that of the crude paper is 0.50.

INCORPORATION OF THREE DIFFERENT RESINS IN PULP SLURRY

W.H. Griggs; U.S. Patent 3,592,731; July 13, 1971; assigned to Eastman Kodak

Company describes a photographic paper base comprising a cationic thermosetting amino-aldehyde wet-strength resin, a cationic thermosetting polyamide-epichlorohydrin wet-strength resin, and an anionic polyacrylamide dry-strength resin. This combination of three different resins gives adequate strength properties while providing an aldehyde concentration which is great enough to provide the desired hardening of photographic emulsion applied to the paper base yet not so great as to adversely affect its sensitometric properties.

The cationic thermosetting amino-aldehyde wet-strength resin is used in an amount of from about 0.001 to about 0.006 part per part by weight of the cellulosic fibers on a dry basis, i.e., about 0.1 to about 0.6% by weight, and more preferably in an amount of from about 0.003 to about 0.005 part per part by weight.

The cationic thermosetting polyamide-epichlorohydrin wet-strength resin is used in an amount of from about 0.003 to about 0.01 part per part by weight of the cellulosic fibers on a dry basis, and more preferably in an amount of from about 0.005 to about 0.008 part per part by weight.

The anionic polyacrylamide dry-strength resin is used in an amount of from about 0.004 to about 0.015 part per part by weight of the cellulosic fibers on a dry basis and more preferably in an amount of from about 0.007 to about 0.012 part per part by weight.

Preferably the amino-aldehyde resin is an amino-formaldehyde resin, or a melamine-formaldehyde resin. The polyamide-epichlorohydrin resin is prepared from a polyalkylene polyamine, adipic acid and epichlorohydrin, or from diethylenetriamine, adipic acid and epichlorohydrin. The polyacrylamide resin is a copolymer consisting of about 85 to about 95% acrylamide and about 5 to about 15% acrylic acid.

While it is essential that the resin system include at least one resin from each of the three classes described, other additives commonly used in papermaking can also be utilized to advantage, and are employed in minor amounts in accordance with conventional practice. Thus, for example, it will ordinarily be desirable to include a water-soluble polyvalent metal salt, preferably aluminum sulfate or aluminum chloride, which aids in setting the anionic polyacrylamide dry-strength resin on the fibers. It will frequently also be desired to incorporate sizing agents in the furnish, preferably sizes derived from stearic acid or related fatty acids.

Water-soluble gums, such as the cellulose ethers, or the cationic starches can also be added to aid in attaining high wet strength. It is however, preferred that the resins be added after the pulp is refined as this gives maximum improvement in the physical properties of the paper.

For optimum results, the order in which the additives utilized are incorporated in the paper furnish should be carefully controlled. As a general rule, all of the anionic materials should be added before any of the cationic materials are added and, in particular, the anionic polyacrylamide dry-strength resin should be added before either of the cationic thermosetting wet-strength resins is added. Moreover, to achieve the desired results the pH should be regulated to attain a value in the range from 4 to 7 and preferably in the range from 4.5 to 5.5.

A preferred procedure is to first refine the pulp to the desired extent, then add the anionic polyacrylamide dry-strength resin, then the sizing agent, then the polyvalent metal salt in an amount sufficient to give the desired pH, then the cationic thermosetting amino-aldehyde and polyamide-epichlorohydrin wet-strength resins, and thereafter form and dry the paper sheet in the conventional manner. After drying, the thermosetting resins should be cured by heating the paper at temperatures of from about 75°C to about 130°C for 1 to 30 minutes.

Example 1: A 3,000 pound mixture of 60% softwood bleached sulfite fibers and 40% hardwood bleached sulfite fibers was dispersed in 6,900 gallons of water containing 15 pounds of carboxymethylcellulose and refined through a double disc refiner and a Jordan.

To one portion of the resulting pulp slurry there was added 0.3% of Resin A, 0.6% of Resin B, and 1.2% of Resin C, $AlCl_3$ in an amount sufficient to give a pH of 4.6, and 2% of sodium stearate sizing agent (prepared by reacting commercial stearic acid with double the stoichiometric proportion of sodium hydroxide), and then a paper sheet, designated herein as sheet 1-(a), was formed in the conventional manner on a papermachine, dried, tub sized with 4% gelatin, redried and lightly calendered.

To a second portion of the same pulp slurry there was added 1.2% of Resin A, $AlCl_3$ in an amount sufficient to give a pH of 4.1, 2% of sodium stearate sizing agent, and 3% of starch, and a second paper sheet, designated herein as sheet 1-(b), was formed in the same manner as sheet 1-(a). For purposes of comparison, the properties of sheets 1-(a) and 1-(b) are summarized below.

Property	Sheet 1-(a)	Sheet 1-(b)
Basis weight (lb/1,000 ft^2)	48.0	48.75
Thickness (mils)	13.8	14.5
Tear strength (Elmendorf), length direction	140	136
Tear strength (Elmendorf), cross direction	180	172
Mullen (psi)	67	58
Wet tensile strength (hectograms per 15 mm width)	47	37

As shown by the above data, sheet 1-(a) exhibits improved strength as compared to sheet 1-(b). Moreover, when both of these sheets were coated with baryta (barium sulfate with gelatin binder) and then with a photographic emulsion the product prepared from sheet 1-(a) was superior to that from sheet 1-(b) as evidenced by improved contrast stability, reduced mottle, and improved image tone stability, resulting from the lower formaldehyde content.

The cationic thermosetting amino-aldehyde wet-strength resin utilized is identified as Resin A and is a melamine-formaldehyde resin which has been dissolved in dilute hydrochloric acid solution and aged to form the acid colloid. The cationic thermosetting polyamide-epichlorohydrin wet-strength resin is identified as Resin B and is a commercially available resin prepared from diethylenetriamine, adipic acid and epichlorohydrin and marketed under the trademarks Kymene 557 and Kymene 709.

The anionic polyacrylamide dry-strength resin is identified as Resin C; it is a copolymer consisting of about 90 weight percent acrylamide and about 10 weight percent acrylic acid and having a Brookfield viscosity in 10% aqueous solution using a No. 4 spindle at 100 revolutions per minute, of 1,250 ± 750 centipoises, and it is marketed under the trademark Accostrength 86. The amounts of wet- or dry-strength resin or other additive are reported in weight percent of the additive based on bone dry fiber weight.

The physical characteristics of the paper were conducted in accordance with standard TAPPI test procedures the identification numbers of which are as follows.

Property	TAPPI Test No.
Basic weight	T-410
Thickness	T-411
Elmendorf tear strength	T-414
Mullen	T-403
Wet tensile strength	T-456

GRAY OPAQUE PAPERS

According to a process developed by *C.R. Beckner and K.B. Kasper; U.S. Patent 3,758,376; September 11, 1973; assigned to Schoeller Technical Papers, Inc.,* gray opaque papers suitable for use in the photographic art are prepared by adding to a suspension of fibers, prior to forming, a light absorbing colorant, titanium pigment particles and a small amount of colloidal material carrying a surface electrical charges.

In such papers, smaller than usual amounts of the light absorbing material are necessary to produce opacity to actinic light. Such papers can be basecoated with a white pigment coating, coated with a diffusion transfer positive emulsion and backcoated with a curl, water and solvent resistant coating to produce a positive print paper for use in pod development in diffusion type photographic processes.

The light absorbing colorants useful in this process were found to be water-dispersible black pigments or water-soluble black dyes. The preferred colorant is a colloidal carbon black pigment. As furnished by the manufacturer, this colorant was a 38% solids colloidal dispersion of carbon black Aquablak B. It is designated as C.I. Pigment Black 7 (C.I. 77266) in the 1956 Color Index.

Water-soluble black dyes have been substituted in whole or in part for the carbon black pigment. A preferred such dye is Nigrosine WSB. It is designated as C.I. Acid Black 2 (C.I. 50420) in the 1956 Color Index.

The titanium pigments include both rutile and anastase titanium dioxides and potassium titanate. Mixtures of titanium dioxide and potassium titanate have been found to be particularly useful. The titanium dioxide pigments are those normally used as fillers in paper and generally have an average spherical diameter in the order of 350 millimicrons.

The potassium titanate is fibrous in character and pigmentary size. A particular fibrous titanate pigment found to be useful had an average particle size 200 millimicrons in diameter and 10 microns in length (sold by E.I. du Pont de Nemours and Company, under the trade name PKT, pigmentary potassium titanate).

At least 10% of the titanium pigment may be replaced by known inexpensive titanium pigment extenders, such as a sodium silico aluminate pigment without noticeably changing the properties of the opaque paper produced.

The preferred amounts and usable ranges of opacifying materials per square meter of paper found to be satisfactory, irrespective of the basis weight or thickness of the paper are shown below. In determining the amounts necessary, a practical range of papers from 5 to 9 mils thick was made and tested.

	Preferred Range	Usable Range
Titanium pigment, grams	14 - 17	14 - 26
Light absorbing colorant, grams	2.5	1.3 - 3.8

The preferred electrically charged colloidal material is a hydrous magnesium silicate consisting of rod shaped particles of an average effective spherical diameter of approximately 20 millimicrons and carrying a negative electrical charge on the particle surface, Avibest C.

Another electrically charged colloidal material which has been found to be useful is a hydrous aluminum oxide consisting of particles having an average diameter of approximately 30 millimicrons and carrying a positive electrical charge on the particle surface, Alon C.

Addition of these electrically charged colloidal materials at the rate of 1.5 to 15% based on the weight of titanium pigment has been found to be necessary to produce an opaque paper using the above ranges of opacifying materials, light absorbing colorants and light scattering pigments.

Opacity Test: A paper is defined as being opaque if the paper will not transmit enough light to visibly desensitize a highly sensitive negative emulsion on exposure to light within a given test time. In the performance of this test, the emulsion side of a high speed negative, such as Polaroid 3000 Speed Negative, is covered with the test paper, placed 9 inches from a 375 watt floodlamp and exposed to the light from the lamp for a period of one minute.

The thus exposed negative is then kept in the dark until developed and fixed using procedures normal for this type of negative. The test paper is said to be completely opaque if the negative is not visibly darkened and black specks are substantially eliminated.

Example: Aqueous dispersions of 2% hydrous magnesium silicate (Avibest) and 50% filler pigments (rutile titanium dioxide pigment containing 10% sodium silico aluminate extender pigment based on the total pigment content) were separately made using a high speed mixer such as a Cowles Dispersator.

The titanium dioxide dispersion additionally contained 0.3% sodium hexametaphosphate based on the titanium dioxide content to assist in dispersing the pigment particles. The magnesium silicate and titanium dioxide dispersions prepared above were added, in that order, to a 3% suspension of photographic grade sulfite and kraft wood fibers which had been previously refined to a Canadian Freeness of about 350 in amounts equivalent to 0.5% magnesium silicate and 10% combined filler pigments by weight based on fiber content.

Prior to formation of the opaque paper on a Fourdrinier machine, 2% cationic starch and about 1% rosin size, both based on the fiber content, and alum to pH of about 4.5 were added to the above combined pulp suspension. 1.5% predispersed carbon black, based on fiber content, was added just after the addition of the cationic starch. The paper web was formed using papermaking techniques known to produce a well-formed paper free from fiber clots and holes, and was pressed and dried in the usual manner. The finished paper was tub sized with 6% oxidized starch and machine calendered to provide a smooth paper for further coating.

The resultant gray paper had a caliper of 7.5 mils; a basis weight of 170 grams per square meter and contained 15.5 grams of filler and 2.1 grams of carbon black per square meter, based on the amount of these materials added. Retention of the filler and other additives was substantially complete since the "whitewater" was recirculated. The paper produced was of a uniform gray color and was found to be completely opaque as determined by the aforementioned opacity test.

A comparative opacity test was made between the paper prepared as described above, designated (A) and three papers made in the manner of (A) in which the hydrous aluminum silicate was omitted (B), the titanium pigment was omitted (C), and the carbon black was omitted (D). The results was as follows: (A), negative clear, no black specks; (B), with no hydrous magnesium silicate, many black specks; (C), with no titanium dioxide, many black specks; and (D), with no carbon black, negative darkened.

PAPER BASE CONTAINING SYNTHETIC FIBER, CMC AND PIGMENT

W.T. Driscoll; U.S. Patent 3,730,830; May 1, 1973; assigned to Eastman Kodak Company found that if an inorganic pigment or carbon is added to the slurry containing carboxymethylcellulose and synthetic polymer fibers, a consistency of the order of 3% can be employed to obtain a uniform dispersion which can be blended satisfactorily with cellulose fibers for forming a paper sheet as usual.

It is possible by this process to make very thin papers having good formations containing about 10 to 25% synthetic fiber and about 1.5 mils thick. By using the synthetic polymers in the process it is possible, for example, to make papers weighing about 3 to 20 pounds per 1,000 square feet having the desired dry and wet strengths usually only obtainable in pure cellulose papers of about 30 pounds per 1,000 square feet containing conventional sizing materials. Preferably a slurry of the polymer fiber, carboxymethylcellulose and the pigment in water is prepared and cellulose fiber which has been refined in known manner is blended therewith in the desired proportion.

The carboxymethylcellulose used in the process contains about 0.4 to 0.8 carboxyl unit per anhydroglucose unit. The carboxymethylcellulose can be used in the form of the free acid, alkali metal or ammonium salts. Useful results are obtained using about 1 to 20%, preferably about 4 to 6%, by weight of carboxymethylcellulose based on the amount of polymer fiber present.

The inorganic pigments used in preparing the paper furnish are particulate materials, preferably those which are photographically inert, i.e., having no adverse effect upon silver halide emulsions coated upon the paper. These pigments include light-colored pigments such as titanium oxide (both rutile and anatase forms), zinc sulfide, barium sulfate, calcium carbonate, etc. In those cases where nonphotographic use of the paper is intended, zinc oxide, lead oxide, etc. can be used effectively to facilitate dispersion of the polymer fibers in the slurry.

Water- or solvent-dispersible types of clays are very useful alone or with the above pigments especially titanium oxide. Sufficient pigment or carbon is used to prevent agglomeration of the polymer fiber in the slurry, e.g., about 15 to 25% by weight of pigment based on the amount of synthetic polymer present.

The proportion of polymer fiber to cellulose fiber can be varied depending upon the feel, wet and dry strengths desired in the paper. For example, 10 to 25% polymer fiber is useful. Representative papers are prepared containing about 2 to 5% retained titanium oxide and about 20% linear polyester, the carboxymethylcellulose largely washing out of the paper when sheeting out the paper on the Fourdrinier as usual.

A wide variety of synthetic polymer staple fibers can be used, preferably those which are water repellant, for example, the well-known textile fibers comprising polyamides, acrylic polymers, modacrylic polymers, acrylonitrile copolymers, cellulose organic acid esters. The polymer fibers can vary in size, e.g., 0.75 to 10, preferably 1.5 to 3 denier being useful in making the paper. Fiber lengths of about 3/16 to one inch can be used as desired.

The papers prepared by the process are especially useful as photographic paper bases. For example, they can be coated with one or more layers of gelatin-silver halide emulsion. Very thin papers prepared by the process are suitable for emulsion coating and provide sensitive papers having good dry and wet strengths in the various photographic processing solutions used. Photo-recording papers advantageously contain a thin paper base made by the process.

Example 1: (a) Preparation of Furnish – To 300 gallons of 30° to 70°F water (if desirable, adjusted to pH 2.0 to 3.0 with sulfuric acid) is added 28.5 pounds of polyethylene terephthalate fiber, 1.5 to 3.0 denier and about 1/4 inch long. To this is added with stirring 5.7 pounds of rutile, medium oil dispersible titanium dioxide and 0.57 pound carboxymethylcellulose averaging 0.6 carboxyl unit per anhydroglucose unit. After thorough mixing, additional 30° to 70°F water may be added to a convenient volume to facilitate flow.

(b) Preparation of Pulp – An 80/20 mixture of bleached softwood and bleached hardwood kraft pulps is defibered in a hydropulper and refined at 3% consistency to a 40 to 50 seconds Williams slowness and pumped to a holding chest.

(c) Use – Part (a) is further diluted 50/50 with water (60° to 110°F) while being pumped to the papermachine head tank where it is mixed with Part (b) the refined cellulose. The flow rates of the two fiber systems are controlled to give the desired ratio of cellulose to synthetic fiber to obtain about 10 to 25% synthetic fiber in the sheet.

The paper is formed, dried and calendered to the desired thickness, e.g., about 1.5 mils. Paper weights of about 3 to 30 pounds or more per 1,000 square feet are obtainable and the paper may contain about 2 to 5% of pigment such as titanium oxide, the carboxymethylcellulose largely washing out of the sheet. The paper has good wet and dry strength. The excellent formation of the paper indicates that the synthetic polymer has been uniformly dispersed in the paper furnish.

NUCLEATED RECEIVING SHEET FOR USE IN PHOTOGRAPHIC CHEMICAL TRANSFER PROCESS

D.J. Beavers and H.C. Yutzy; U.S. Patent 3,518,160; June 30, 1970; assigned to Eastman Kodak Company provide a receiving sheet paper for use in the photographic chemical transfer process by immersing the freshly made paper web through at least one sizing bath and then passing it through press rolls to force the sizing into the stock and to remove the excess from the surface. The size comprises a hydrophilic colloid and silver precipitating nuclei and preferably an antistain agent.

The antistain agent is preferably a slightly soluble or water-insoluble acid such as boric, phthalic, benzoic, adipic, etc., or ammonium salts of more water-soluble acids. Solvents for these acids may be used to incorporate the acids into the raw paper stock, but it is preferable to add the acids to the nuclei gel size as slurries, dispersions, or hot solutions. The amount of antistain agent will depend upon the paper thickness but is preferably within the amount of 0.025 to 1.2 grams per square foot. A further restriction on the use of low solubility acids is the maximum solution temperature of about 120°F.

It has been found that, when an organic acid (such as phthalic acid) has borderline solubility at the desired temperature and concentration, the addition of a small amount of another organic acid (such as benzoic acid) of lower solubility will result in a combined solubility which is greater than either acid separately. Thus a mixture of 82.5% phthalic acid and 17.5% benzoic acid can be dissolved (and remain dissolved) at 3% solids in water at temperatures of 100° to 120°F.

The nuclei, which may be conventional nuclei types, such as colloidal metals or metal sulfides, e.g., colloidal silver, silver and zinc sulfides or various nuclei mixtures, may be applied within the range of 0.1 to 50 milligrams per square foot for the first size bath. These nuclei preferably have an average diameter of about 25 to 2500 A.

In the second size bath no antistain agent is used, but only nuclei and a binder. The concentration of the nuclei is such that an amount within the range of 0.1 to 75 milligrams per square foot is applied on the paper stock. These quantities may be varied, but the total coverage is preferably within the range of 1 to 10 milligrams per square foot but may be 0.1 to 100 milligrams per square foot.

The paper base should be tub sized with solutions more concentrated than the normal gelatin size of two or four percent. Best results are obtained with gelatin solutions of six percent or more, preferably about ten percent. Two tub sizes are better than one tub size since the paper base has a smoother surface and gives prints with little or no mottle.

Example 1: Preparation of 0.02 N ZnS Nuclei, 12% Gel (6-Liter Batch) – To the kettle at 40°C, add 2,880 grams 25% gelatin solution, 1,270 ml H_2O, a solution of 318 grams sodium thiosulfate ($5H_2O$), 15 grams sodium sulfite and 12 ml gel preservative. Dissolve and add 120 ml of 1.0 N sodium sulfide and stir 3 min.

A solution of 125 ml zinc nitrate (1.0 N) is diluted to 1,580 ml with water and this solution is added to the kettle in 5 minutes at 40°C. This nuclei gel melt is used as a tub size. An antistain agent may be added, and if desired, nuclei cold toners.

Example 2: (a) First Tub Size – To 300 grams of 0.02 N ZnS nuclei (12% gel) melted at 50°C is added a solution of 85 grams phthalic acid in 500 ml methanol over a 5 minute period. This turbid solution used as the antistain first tub adds about 0.6 gram per square foot of solids to non-tub-sized stock.

(b) Second Tub Size – To 6,000 grams of 0.02 N ZnS nuclei in a 10 to 12% gel melt was added a solution of cold toning agent. Using the first tub size plus this second tub sizing adds about 0.9 to 1.1 grams per square foot of solids to non-tub-sized paper. Excellent nonyellowing two-sided chemical transfer positive prints are obtained when an exposed silver chlorobromide emulsion is processed with a silver halide developer in the presence of a silver halide solvent.

Example 3: Nucleated Receiver Sheet – A tub sized nucleated-acid-treated receiving sheet was prepared by passing a paper support through two consecutive tubs containing the following ingredients: first tub, 12% gelatin solution containing 3,500 grams of water, 85 grams of phthalic acid and 0.02 N zinc sulfide; second tub, 12% gelatin solution containing 0.02 N zinc sulfide.

A second nucleated receiving sheet was prepared by surface coating a gel sized paper support on both sides with a 7% gelatin solution containing 0.02 N zinc sulfide. A negative sheet was prepared by coating a paper support with a silver chloride emulsion at a coverage of 90 milligrams of silver and 220 milligrams of gelatin per square foot. Over the emulsion layer, there was coated a layer of carboxymethylcellulose at a coverage of 100 milligrams per square foot.

The negative was exposed to a line subject and bathed in a diffusion transfer developer of the type described in Example 4 of U.S. Patent 2,987,396. During development, the exposed negative sheet was squeegeed in contact with the tub sized receiving sheet described above for 10 seconds and the two sheets then separated.

A similar exposed negative was squeegeed in contact with the receiving sheet having the nuclei coated on the surface. After a 10 second transfer time, this sheet produced an image having considerably lower D_{max} and poorer quality than when the tub-sized nucleated receiving sheet was used.

Example 4: Sized, Nucleated Chemical Transfer Sheets Used as Receiver Sheets in a Normal Solvent Diffusion Transfer System – (a) A sized, nucleated sheet was prepared on a tub sizing machine in the following manner. The first tub contained 0.02 N ZnS, 12% gelatin and hypo while the second tub contained, in addition, an appropriate cold toner. This nucleated, sized sheet was overcoated with an interlayer of 0.02 gram per square foot of ethylcellulose phthalate to facilitate the wash-off of the silver chlorobromide emulsion-ethylcellulose phthalate layer which was coated over the interlayer.

(b) A nucleated, sized receiver sheet prepared as in (a) and a similar sheet which contained, in addition, an antistain agent were used as receiver sheets in a solvent-diffusion transfer system. A coated sample of a high speed silver bromoiodide emulsion was exposed on an Eastman IB sensitometer with an appropriate filter for 1/25 second.

The exposed sample was then processed by attaching a pod containing a viscous solvent silver halide developer between the exposed negative element and the respective receiver sheet and passing them, in sandwich form, between two rollers. After processing times of 10 seconds and 20 seconds, the negative was stripped from the receiving sheet and the positive image thereon was stabilized.

All the positive images were found to have a sharp, dark brown image of very high quality and density. A high density, cold-black image would be obtained in place of the brown image by the incorporation of an appropriate cold toner in the system.

HIGHLY STABLE RESIN COATED PAPER

I.H. Crawford, R.E. Democh and R.J. Baron; U.S. Patent 3,755,069; August 28, 1973; assigned to Eastman Kodak Company provide a resin coated paper product, suitable for use in photographic applications, comprising a paper core containing a resin stabilizer and/or antioxidant capable of migrating into a resinous overlayer coated onto the paper core.

According to one aspect of the process, a suitable antioxidant or stabilizer, or mixture of same is coated onto the surface of a paper stock, dried and subsequently overcoated with a layer of the polyolefin. The stabilizer migrates into the polyolefin layer from the substratum of stabilizer when the coated paper product is subjected to conditions of normal use.

Preferably, the antioxidant, stabilizer or mixture is incorporated directly into the paper stock, during the papermaking process, and the paper thus formed coated with the polyolefin resin. In this configuration the stabilizer migrates from the paper core or base into the polyolefin layer to provide a coated photographic paper which exhibits superior stability while still maintaining its desired wet-strength, resin-to-paper bonding and other valuable characteristics.

The processes are particularly applicable to the coating of paper base with polyolefin resins, particularly polyethylene. Papers suitable for use include almost any conventional paper product susceptible to resin coating according to well-known techniques. Particularly preferred in photographic applications are the baryta-coated photographic stocks.

Resin stabilizers can be broken down into the following five main chemical groups based upon the mechanism of their activity: (1) chain terminators, (2) chain terminator regenerators, (3) peroxide decomposers, (4) metal deactivators and (5) light absorbers (UV and energy transfer agents).

Specific classes of chemicals that fall into each of these categories are as follows: (1) phenols, bisphenols, thiobisphenols and amines, (2) certain sulfur containing compounds, (3) polymeric phosphorus compounds, sulfur and phosphorus compounds, (4) sulfur compounds, polymeric phosphorus compounds and chelating agents and (5) benzophenones, salts of salicylic acid, benzotriazoles and metal organics.

Specifically preferred as the stabilizer in the poly(α-olefin) coated papers are those contained in groups, 1, 3, 4 and 5 listed above. Of particular use as stabilizers for the polyolefin resins are the following specific compounds and mixtures thereof: octadecyl(3,5-di-t-butyl-4-hydroxyphenyl) acetate, (3,5 di-t-butyl-4-hydroxyphenyl) orthophosphate, 4,4'-thiobis(6-t-butyl-meta-cresol), 2-benzoimidazolethiol, 2-naphthyl disulfide, n-dodecyl sulfide, 2-naphthalene-thiol, tricetyl trithio phosphite, and zinc(II) dibutyl dithiocarbamate.

Example 1: Antioxidant Coating – Grade 504 paper stock is coated with a 2% by weight solution in toluene of these resin antioxidants, used individually:
(1) Irganox 1076, octadecyl (3,5-ditertiary-butyl-4-hydroxyphenyl) acetate
(2) Ethyl 796, tris(3,5-ditertiary-butyl-4-hydroxyphenyl) phosphate.

The dry antioxidant coverage was 0.5 gram per square meter. Comparisons are prepared by coating with solvent alone, and uncoated paper is included. All samples are then extrusion coated with a medium density polyethylene (0.935, melt index of 8.0, ASTM D 1238, condition E) containing 10% titanium dioxide, 0.85% colorants and 0.1% stearyl stearamide.

Total polyethylene coverage is 7.5 pounds per 1,000 square feet. The samples are aged in ovens, with and without visible light illumination. Samples with the antioxidant coatings treated in fashion described below show a marked improvement in stability and resistance to yellowing over the comparisons.

To test the stability of samples, they are incubated at 105°C for 72 hours and percent reflectance measured at 430 nanometers. Samples containing the antioxidant incorporated as described demonstrate a 5% loss in reflectance while nonantioxidant coated samples demonstrate a 10% loss.

In another test similar samples are exposed to a fluorescent light source at an ambient temperature of 90°C until a loss of strength occurs. The antioxidant treated resin coated papers require 184 hours to fail, while the untreated papers fail at 112 hours, a stability improvement of 64%.

Example 2: Stabilizer Added to Stock – Handsheets are made by dispersing a four stabilizer mixture of:

40% Butyl Zimate [zinc(II) dibutyl thiocarbamate],
15% Ethyl 330 [1,3,5-trimethyl-2,4,6-tris(3,5 di-tertiary-butyl-4-hydroxytolyl)-benzene],
30% CAO-6 [2,2'-thio-bis(4-methyl-6-t-butylphenol)],
15% benzimidazolethiol.

in stearate size with a Waring blender. The dispersion is added to the pulp and precipitated with aluminum chloride. Handsheets are made at 38 pounds per 1,000 square feet with stabilizer added at 0.25, 0.5, 1, 2 and 4% levels based on the pulp weight. The sheets are extruded with 7.5 pounds per 1,000 square feet of a medium density, titanium dioxide pigmented polyethylene on the face side and 8.0 pounds per 1,000 square feet of resin, a clear high density polyethylene (0.950 melt index of 7.5), on the back side. Results of testing as described above show a marked improvement in resin stability.

Example 3: Stabilizer Added With Surface Size – A surface sizing solution is prepared by cooking Claro 5583 starch made by Corn Products Co. at 3.5% solids by weight and adding 8% Na_2SO_4. 2-naphthyl disulfide at 1% by weight is dispersed in this solution by dispersing it in a Waring blender in the presence of 0.2% ammonia salt of polymethacrylic acid as surfactant and 0.2% Kelzan, a high molecular weight polysaccharide, as a suspending aid.

The resulting surface sizing solution is tub sized using a non-tub-sized 38 pounds per 1,000 square foot paper stock. The paper is then calendered. No difference is physical properties is observed between the 2-naphthyl disulfide sized papers and papers sized without the 2-naphthyl disulfide in the size solution.

The papers are extruded with polyethylene with excellent adhesion and are sensitized with a gelatin silver halide emulsion. 2-naphthyl disulfide pickups based on the weight of the paper substrate are 0, 0.21 and 0.28%. These papers demonstrate improved stability.

EXTENSIBLE PAPERS

IMPROVING TENSILE STRENGTH BY EXTERNAL STARCH SIZING

Extensible (compacted) paper produced, for example, in accordance with the apparatus and process disclosed in U.S. Patent 2,624,245 has certain well-recognized advantages and commercial uses. Such paper is subjected, while in a partially moistened condition, to compressive compaction in the direction of web movement (machine direction or MD) between a pressure nip, thus compacting and forcing the fibers together to produce an inherent stretchability without creping. Compacted paper has improved tensile energy absorption (TEA), burst and tear characteristics which are highly desirable for such end uses as the manufacture of paper sacks. However, extensible paper has reduced tensile strength and stiffness, the diminution of the latter characteristic increasing the difficulty of converting this paper into sacks.

A. Heim and E.J. Groome; U.S. Patent 3,908,071; September 23, 1975; assigned to Clupak, Inc. are concerned with improving the tensile strength and stiffness of compacted paper without sacrificing tear strength and additionally, to produce a radical improvement in TEA and burst strengths. This is achieved by the selective impregnation of starch as an external sizing, particularly in a manner which will not destroy the extensible nature of the paper.

It has been discovered that the controlled application of a water solution containing starch to both sides of a compacted paper web under optimum conditions will increase the tensile strength of the web by 150%, burst strength by 400%, tensile energy absorption by 300% and stiffness by 100%. As a result of these drastic and unexpected improvements it is possible to utilize some pulp stock (furnishes) previously thought to be totally unsuitable for the manufacture of paper for sacks, and alternatively makes possible the use of less, conventional raw pulp stock without sacrificing strength. Under identical conditions greater improvements in tensile strength and stiffness are obtained with cornstarch rather than potato starch.

It is suggested that slightly converted (plasticized) starch should be superior to highly converted starches, particularly at high percentage levels of starch impregnation. Such plasticized starch should be highly compatible with the extensible qualities of compacted paper. This is not, however, in any way to suggest that the use of unconverted starch is undesirable, since the examples disclosed herein were made using starch in the highly converted form. The process is described with reference to Figure 8.1.

FIGURE 8.1: IMPROVING TENSILE STRENGTH OF EXTENSIBLE PAPER BY EXTERNAL SIZING

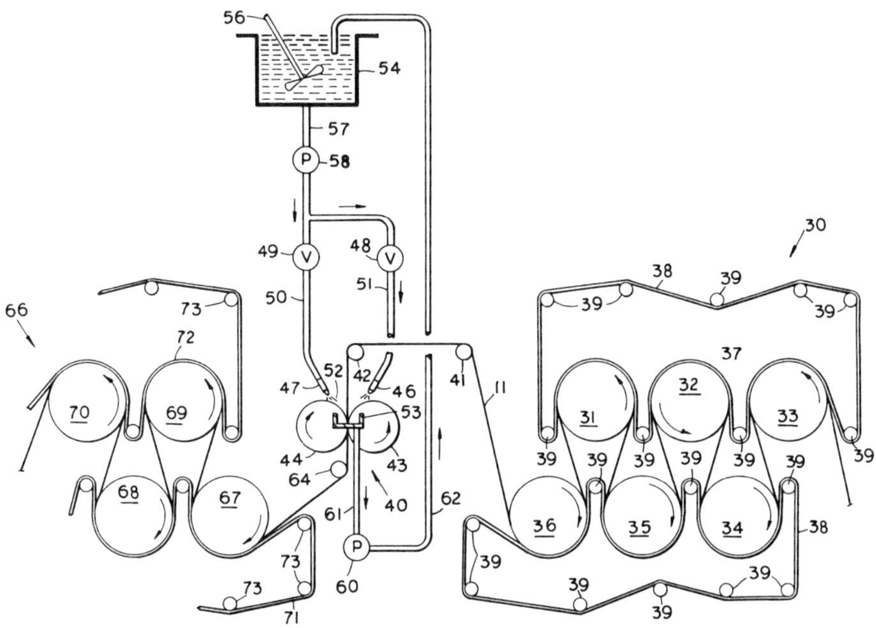

Source: U.S. Patent 3,908,071

After leaving the compacting apparatus, the web **11** will pass through the dryer section **30** which includes conventional drying rolls **31-36**, the web **11** first passing over drying roll **33** thence to roll **34**, etc., as shown. Felt dryer belts **37** and **38** are arranged upon guide rolls **39** and in contact with the respective dryer rolls **31-36** to assure the intimate engagement between the web **11** and the respective dryer rolls.

After leaving the compactor, the moisture contained in the web is approximately 35 to 45% moisture by weight. After passing through dryer section **30**, the moisture content of the web **11** will be approximately 7 to 15% of moisture by weight.

Reference numeral **40** indicates a size press arrangement for treating the dried compacted web **11** in accordance with this process. This web first passes over guide rolls **41, 42** and therefrom proceeds between a nip formed by two counterrotating size press rolls **43** and **44**. Nozzles **46, 47** control the starch metering valves **48, 49** in conduit **50, 51** which permit a metered quantity of a starch solution **52** to flow into the nip between size press rolls **43, 44**. Edge catch pans **53** collect the excess starch solution at each end of the nip. The starch solution is supplied from a starch solution supply tank **54** having therein an agitator **56**, starch in solution being drawn from the tank **54** through conduit **57** to the pump **58** which supplies metering valves **48** and **49**. A starch return pump **60** serviced by line **61** from catch pans **53** returns the starch solution via line **62** to the supply tank **54**.

During the passage of web **11** between the nip of press roll **43** and **44**, a solution of starch is applied under pressure to both sides of the web in accordance with varying percentages indicated below by way of specific examples. The passage of the web **11** through the starch solution and between the rolls **43, 44** will be sufficient and will be so regulated as to the speed of application and the strength of the starch solution to permit and to cause impregnation of starch into both surfaces of the web **11** in accordance with a predetermined amount. The weight of starch impregnated and deposited in the web will be from 1 to 10% by weight to the total weight of the web after drying in order to achieve good results. During this process the moisture content of the web will increase from the 7 to 15% value at the outlet of the dryer section **30** to a value of 35 to 45%.

Following the application and impregnation of starch into both sections of the web **11** the web passes around guide roll **64** and thence into the second dryer section **66**. Dryer section **66** contains a plurality of dryer rollers **67, 68, 69, 70**, etc., about which pass dryer felts **71, 72** guided by rolls **73**. The dryer section **66** is similar in all respects to the dryer section **30** and the web **11** passing therethrough will lose the moisture imparted by the size press exiting from the dryer section **66** with moisture content of approximately 5 to 10% by weight.

The difference between the pressed arrangement and that shown in U.S. Patent 3,362,869 (reported elsewhere) is that in the latter the starch applied is a surface coating used to achieve adhesive bonding between the web and the dryer roll, whereas in this method the web becomes impregnated with starch on both sides to achieve the subsequent additional improvements of radically increased tensile, burst and TEA strengths. Such additional effects are made possible by the use of the size press, which includes the size press rolls for forcing the starch solution into both web surfaces. In the following examples, CD refers to cross direction and MD is machine direction.

Example 1: The constituents used in this example were a furnish of unbleached 30% softwood, unbleached 70% hardwood; unrefined stock (12 sec. Williams Freeness); and starch (acid modified cornstarch).

	Starch Added		
	0%	**2.8%**	**5.7%**
Tensile, lb/in:			
MD	6.0	14.0	21.0
CD	5.5	8.8	10.6

(continued)

	Starch Added		
	0%	2.8%	5.7%
Elongation, %:			
MD	9.1	8.1	8.5
CD	1.3	2.3	2.8
TEA, ft lb/ft²:			
MD	4.17	7.95	14.12
CD	0.62	1.66	2.60
Tear, grams:			
MD	54.8	63.6	66.8
CD	61.6	76.5	88.8
Burst, psi	5.4	17.7	23.7
Stiffness, Tabor units:			
MD	0.40	0.98	1.19
CD	0.48	1.01	1.13

Example 2: The constituents used in this example were a furnish of unbleached 30% softwood, unbleached 70% hardwood; refined stock (40 sec. Williams Freeness); and starch (acid modified cornstarch).

	Starch Added			
	0%	1.4%	3.8%	5.7%
Tensile, lb/in:				
MD	14.6	21.7	26.3	29.6
CD	9.0	10.1	12.4	12.6
Elongation, %:				
MD	10.4	9.7	10.9	8.6
CD	2.4	3.4	3.5	3.9
TEA, ft lb/ft²:				
MD	12.7	16.6	21.1	27.2
CD	1.97	2.80	3.87	4.42
Tear, grams:				
MD	78.7	75.2	73.9	68.5
CD	97.6	101.1	121.4	90.9
Burst, psi	18.8	28.6	30.8	35.6
Stiffness, Tabor units:				
MD	0.65	1.16	1.20	1.73
CD	0.85	0.89	1.01	1.31

MACHINE GLAZED EXTENSIBLE PAPER—STARCH ADHESIVE APPLIED TO ONE FACE OF WEB

H.S. Welsh; U.S. Patent 3,362,869; January 9, 1968; assigned to Clupak, Inc. has developed a process for producing a machine-glazed extensible web. This is accomplished by preparing a paper web so that it has extensibility in excess of its primitive extensibility, surface treating the web with a water-thin solution of adhesive without any substantial disturbing or relaxing of the fiber network of the web, and subsequently drying the web so treated on a dryer roll so that a high gloss is imparted to the treated web surface.

The paper web is formed in any conventional manner, for example, a water-laid furnish of cellulose. The web is then partially dried to a coherent structure of 30 to 40% moisture content. The web is then made extensible. Any apparatus such as the Cluett mechanical compaction apparatus would be applicable to impart extensibility to the web. In the Cluett compaction apparatus, a paper with sufficient water removed to create a coherent fibrous structure is submitted to a process which pushes and crowds the fibers of the web together in the space between the opposed faces of the web as laid by forces applied in the direction parallel to the faces of the web and opposed to the direction of its lengthwise movement continuously and uniformly over the web.

Upon subsequent drying by conventional means, the modified orientation of fibers and fibrils is retained in the dry state and the sheet takes on firmness and strength due to the mutual adherence of the finely-entangled structures within the web. A resistance to deformation is produced which allows continuous deformation in the direction of extensibility until rupture occurs. Means are available for imparting extensibility to a web in multiple crossing directions so that continuous resistance to deformation is exhibited in more than one direction. The elongation is characterized by no substantial decrease in thickness of the web.

The extensible web is then passed through an applicator means which applies a water-thin solution of adhesive to one surface of the web. The treated surface is then placed against the polished surface of a machine glazing cylinder before the water has an opportunity to penetrate substantially into the body of the web. The web is removed from the roll after removal of all but a residual amount of moisture and is found to have a high gloss on the surface which was adhered to the cylinder.

Since less total water is present in the web at the point of contact with the glazing cylinder, less drying during the glazing period is required than in conventional commercial operations. Thus the lineal speed of the web can be increased or, in the alternative, a small cylinder can be used without loss of capacity. This is an advantage since larger cylinder diameters necessitate an increase in dryer shell thickness which retards heat transfer and largely nullifies gains realized from larger drying areas. Furthermore, associated problems of fabrication, spacing and shipping make utilization of the smallest possible cylinder desirable.

The process is described with reference to Figure 8.2. This figure shows a mechanical compaction apparatus **1** into which the web **2** is introduced after having been dried to approximately 30 to 40% moisture by weight. The compaction apparatus imparts extensibility to the moving web. Extensibility of 0 to 25% or more can be imparted to the web by a single unit. Further operations on the web may be carried out for even greater extensibility.

The web is moved along suitably mounted rollers **3** and **4**. After compaction, the web has approximately 25 to 35% moisture by weight. The method of imparting extensibility by mechanical compaction is set out in U.S. Patent 2,624,245 and reference can be made to that patent for the technical details of mechanical compaction.

A flexible blade **5** is mounted by mounting means **6** so that it rides in tension against one surface of the moving web. A water-thin solution of adhesive **7** is applied to the trailing surface **8** of the blade which then transports it to the surface of the web and applies it thereon in a uniform fashion. A storage tank **9** is located in proximity to the applicator blade **5**. A pipe or conduit **10** connected to the tank transports the water-thin solution to the upper portion of the trailing surface of the blade.

FIGURE 8.2: MACHINE GLAZED EXTENSIBLE PAPER

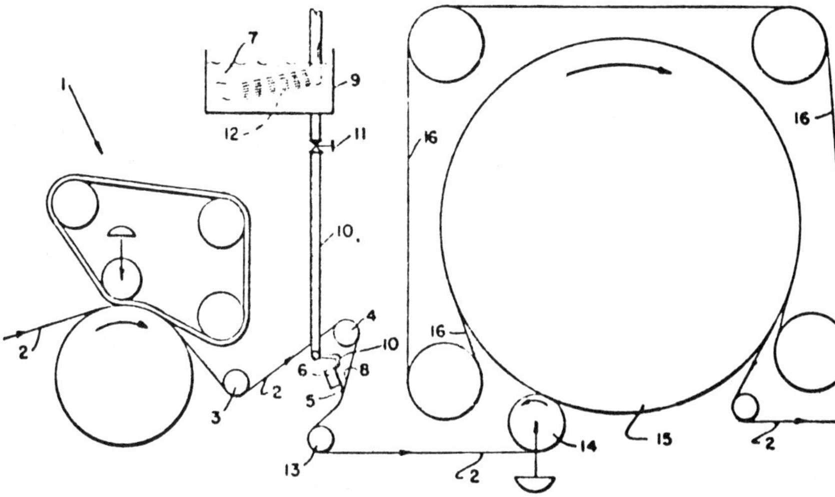

Source: U.S. Patent 3,362,869

A metering valve **11** is disposed in the line from the tank so that the amount of solution applied to the surface of the web per linear foot can be precisely controlled. A steam coil **12** is disposed within the storage tank to provide a means for controlling the temperature of the solution. The temperature is not critical, but it is desirable to heat the solution to reduce the amount of heat required for drying of the paper web.

After being surface treated, the web is moved along by rollers **13** and **14** to a rotatably mounted steam heated glazing cylinder **15**. Roll **14** is a pressure roll. This roll is loaded in a conventional manner so that it applies enough pressure to cause the moving web to adhere to the polished surface of the glazing cylinder. The nip pressure produced by roll **14** and the drum **15** must be sufficient to adhere initially the plastic surface of the web to the polished contour of the glazing cylinder. Pressures in the range of 300 to 400 lb per linear inch have been found satisfactory. A felt backing **16** may be utilized in a conventional manner during the drying process, although its use is optional.

In general the distance between the applicator means and the point of adhesion to the dryer roll should be kept at a minimum. The proper distance in a given operation will depend on absorptivity of the web and the speed of the web.

The glazing cylinder is a large metal drum **15** having a polished chrome plating or other smooth finish on its peripheral surface. A cylinder temperature of about 200°F has been used but the appropriate temperature is easily determined depending on the grade of paper being processed. The cylinder is mounted rotatably.

Example: Pilot Operation — A web of unbleached kraft paper of the sort used in wrappings having approximately 28% moisture by weight was treated with a solution of starch of a mixture of about 5% starch by weight. This represents about 1% of starch when pounds of starch is compared with pounds of paper. A typical water-thin solution would contain approximately 2 to 6% solid content of starch by weight. The Yankee dryer roll was maintained at approximately 215°F and the nip pressure, that is, the pressure produced by the roller and the drum, was about 350 pounds per linear inch.

The resulting gloss was 23 as measured on the L.R.L. Gloss Meter, Serial No. 1. Control samples of paper of about 28% moisture content by weight were run without surface treatment to demonstrate the effectiveness of the method. After drying on the Yankee cylinder, a gloss from 6 to 7 was produced, compared with the gloss of 23 obtained with the method.

The adhesive solution used is water-thin and accordingly has a low viscosity. It has been found that for best results the web should have above approximately 22% moisture by weight when the water-thin solution is applied. The amount of starch in solution can be varied. Starch solutions of approximately 1.4% to approximately 5.5% were successfully utilized to produce a high gloss surface.

USE OF GLYCERIN LUBRICANT TO PREVENT SHRINKAGE OF PAPER WEB

The process of *L. Rabstad; U.S. Patent 3,477,906; November 11, 1969; assigned to M. Peterson & Son A/S, Norway* relates to the operation of a so-called shrinkage press, by means of which web material, particularly paper, is treated in order to acquire a better extensibility. The process is described with reference to Figure 8.3.

FIGURE 8.3: USE OF GLYCERIN LUBRICANT TO PREVENT SHRINKAGE OF PAPER WEB

a.

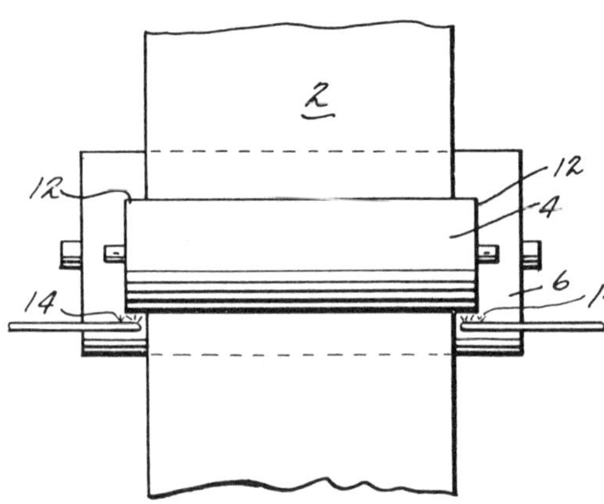

Top Plan View of Apparatus

(continued)

FIGURE 8.3: (continued)

b.

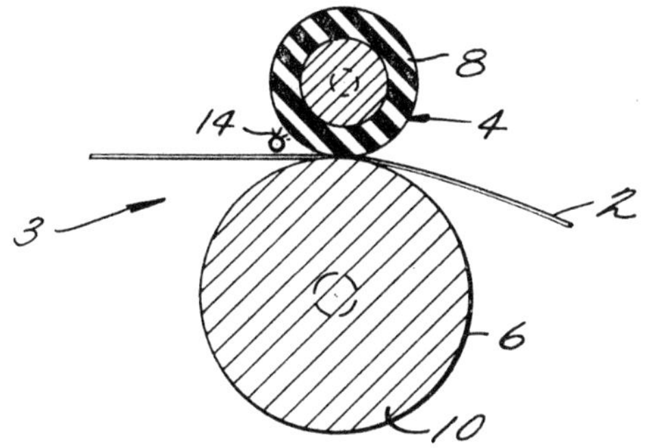

End Cross Section View of Apparatus

Source: U.S. Patent 3,477,906

In the shrinkage of a running paper web in accordance with a known method, the paper web **2** to be treated is caused to pass in a moist condition through a press nip **3** formed by two rolls, **4** and **6**; one roll **4** is provided with a surface layer of an elastic and resilient material **8**, preferably rubber, while the other roll **6** is made with a hard and smooth metallic surface **10**.

The rolls are connected mechanically or electrically in such fixed mutual relation that the hard roll is driven with a peripheral speed about 10 to 20% higher than that of the rubber-clad roll. The rolls are pressed against each other with a line pressure of up to 30 kg/cm and the hard roll **6** is heated to between 50° to 150°C. The rubber layer **8** which is deformed in the press nip **3** contracts on the exit side of the nip and due to the frictional engagement with the paper web **2** the latter is caused to take part in the contraction while sliding against the hard roll **6** which rotates at higher speed. In this manner the web is shrunk in its longitudinal direction without noticeable creping so that the paper, after stabilization of its structure by drying, acquires a higher degree of extensibility than untreated paper.

It is difficult to guide the paper web so that its position always coincides with the press nip accurately. If the edges of the web would reach outside the press nip, which may be caused by a quite normal transverse movement of the web by a few centimeters, its edges will be untreated and therefore not shrunk in the longitudinal direction, but will take on a wavy or crinkled contour making it difficult to wind up the web with the use of conventional equipment for that purpose.

On the other hand, if the paper web is chosen so wide relatively to the length of the press nip that the edges thereof will always pass through the press nip, then at the ends **12** of the nip the rubber-clad roll **4** will be pressed directly against the more rapidly rotating heated hard roll **6**. On account of the heavy friction against the hard roll and the heat transferred therefrom, the rubber layer will be subjected to heavy stresses requiring that the material used therefor has high wearing resistance, high heat permanence and low rate of decomposition in the surrounding atmosphere. If all the requirements cannot be satisfied, the wear between the rolls at the contact points will cause wear, and as lateral movements of the web are almost unavoidable, wrinkles will soon occur at the edges of the web. Therefore, the soft roll will have a comparatively short life and must be replaced, which involves work stops and costs.

By supplying a lubricant **14** to those points of the press nip where the treated paper web does not separate the surface of the elastic roll from that of the opposed hard roll, the wear is practically eliminated at the places where the rolls directly contact each other. Therefore, the paper web may very well be of a somewhat smaller width than the length of the common generatrix of the rolls and nevertheless a durable shrinkage press can be obtained without a too restricted choice of materials for the surface layers of the rolls rubbing against each other.

The lubricant should be inert and of high viscosity. A lubricant which has proved highly satisfactory is glycerin. The lubricant may be supplied by letting it drip onto or by spraying it upon one roll or both rolls at the ends thereof that come into direct contact with each other.

The process is applicable not only to a shrinking press of the kind mentioned above and known, e.g., from U.S. Patent 3,104,197, but also to a shrinking press similar to that shown in U.S. Patent 2,624,245 and in which the elastic layer consists of an endless blanket running in a loop and pressed by means of a sustaining roll or any other sustaining member against a heated hard roll.

APPARATUS FOR MECHANICALLY COMPACTING A CONTINUOUS WEB TO EFFECT STRETCHING OR SHRINKING

A. Schmidt; U.S. Patent 3,471,363; October 7, 1969 describes a process for stretching or shrinking a continuous web including the steps of pressing the web against a flexible deformable band, moving the band together with the continuous web, successively increasing or decreasing, in step-wise manner, the pressure exerted on the band, thus causing the band to expand or contract, respectively, in a direction transverse to the direction of travel and causing a corresponding stretching or shrinking of the continuous web in the same direction.

An apparatus for effecting this process includes a pressure cabinet, one wall of which is formed by a flexible deformable band, the band being designed for deformation in a direction transverse to the path of travel of the continuous web, but not in the direction coinciding with the path of travel. The pressure cabinet includes a number of separate pressure chambers and a pressure control system which provides for successively higher or lower pressure in each pressure chamber.

EXTENSIBILITY IN MACHINE AS WELL AS CROSS MACHINE DIRECTION

The process of *C.B.A. Ihrman; U.S. Patent 3,523,865; August 11, 1970; assigned to Billeruds Aktiebolag, Sweden* relates to a method of treating a running web of paper in such a manner as to shrink the paper web in the machine direction as well as in the cross machine direction.

This shrinkage is achieved by the combination of a mechanical shrinkage unit, which shrinks the paper web in the machine direction, with a second shrinkage unit involving removing moisture from the web, or drying the web, while the paper web is supported by a heated gaseous medium, e.g., air, carbon dioxide, nitrogen, etc., whereby shrinkage of the web in its cross machine direction is obtained.

The process allows the production of extensible paper with the desired degree of extensibility in the machine direction as well as the cross machine direction, at high machine speeds. The process is described with reference to Figure 8.4.

In Figure 8.4a, **A** designates the conventional wire-forming and wet press section of a papermaking machine. For example, a slurry containing 0.5% of a cellulosic fiber material and 99.5% of water is supplied to this section, and a running paper web is formed therein and dewatered to a moisture content of about 70%. B_1 designates a predrying section, comprising a series of conventional steam heated rolls, against which the delicate paper web is held in contact so that a part of its water content is evaporated.

The web leaves this section with a moisture content of about 50% and enters the next section, e.g., the air-drying section B_2, in which the shrinking unit may be incorporated and which is shown in more detail in Figure 8.4b. With a moisture content of about 15%, the web then runs into the final drying section B_3 comprising a series of conventional steam-heated rolls and in which the moisture content of the web is reduced from about 15% to 5 to 7%, i.e., the air-dry value at which the moisture is at equilibrium and the web neither absorbs moisture from nor gives off moisture to dry atmospheric air. By performing the final drying with the paper being supported upon the surface of cylinders, cockling will be prevented. The paper web is then calendered in section **C** and rolled up.

The air-drying section B_2 comprises a number of blow boxes **1** arranged in five tiers, each tier consisting of horizontal parallel boxes located at the same level and having a width, measured in the cross machine direction of the papermaking machine, essentially corresponding to the width of the paper web. Heated air or other heated gaseous medium is supplied to both ends of each blow box and is let out as jets **2** preferably directed alternately in the forward and backward direction, as related to the running direction of the web, through eyelid openings **3** provided in the top wall of the blow box and evenly distributed over its length (Figure 8.4c).

The web **w** is led back and forth over the tiers of blow boxes in five superimposed passes or runs **11, 12, 13, 14** and **15** separated by turning rolls **16, 17, 18** and **19** located at opposite ends of the drying section B_2. The air jets form cushions of air upon which the paper web floats freely, or is supported, so that its weight is counterbalanced and need not be supported by the turning rolls—in other words, the web is airborne.

Due to interaction between the static and the dynamic air pressure upon the web, the paper web will be held at a constant distance, determined essentially by the velocity of the air, from the tops of the blow boxes. An overpressure in the boxes on the order of 1" water column has proven to give stable running conditions.

FIGURE 8.4: EXTENSIBILITY IN MACHINE AS WELL AS CROSS MACHINE DIRECTION

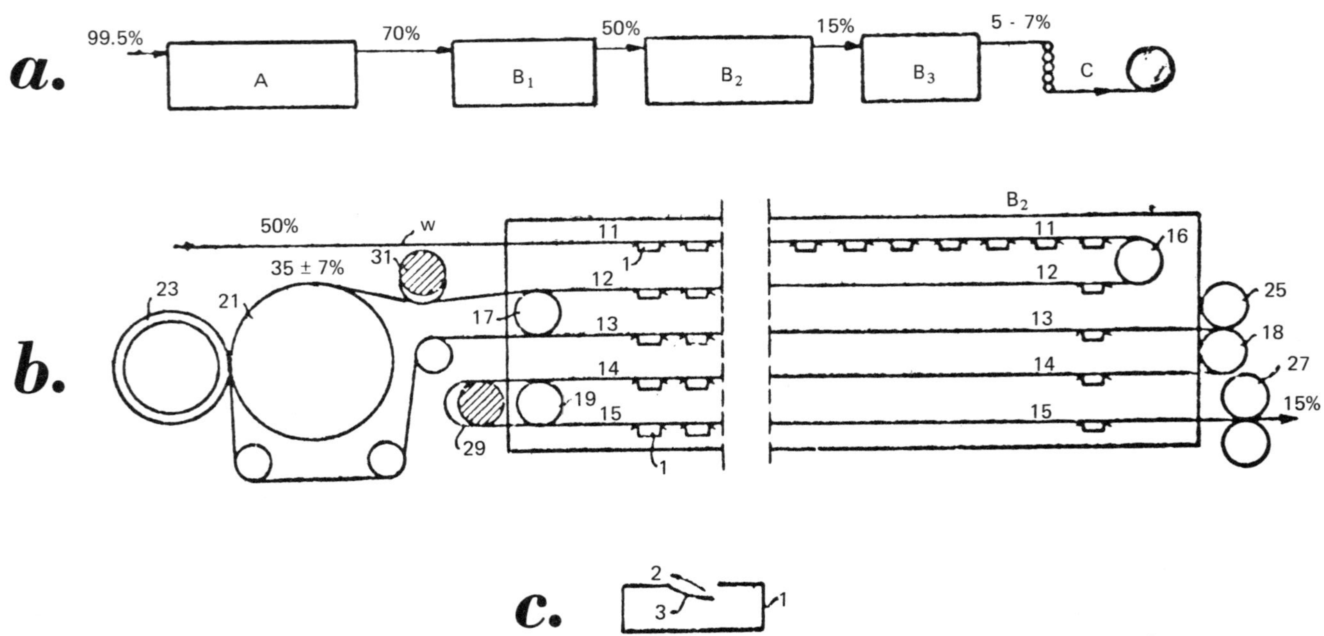

(a) Block Diagram of Papermaking Machine
(b) Air Drying and Shrinkage Section of Machine
(c) Cross Section of Blow Box

Source: U.S. Patent 3,523,865

There is only a very slight friction between the web and the air cushions and therefore the paper web can be advanced with a very small force in its machine direction. Such a force, which is called the draw tension, is insignificant in comparison to the forces in the paper striving to contract or shrink the paper as it dries on account of evaporation of part of its moisture content caused by the heating by the air jets.

In the cross machine direction the web is fully or almost fully unrestricted to shrink, but in the machine direction the shrinkage is restricted more or less according to the draw force or tension applied. This draw force is preferably chosen above the value required for merely overcoming the friction opposing the movement of the web, for the purpose of maintaining the web smoothly extended and properly guided, so that it does not drift sidewise.

While flowing along the under side of the web the heated air absorbs moisture therefrom. The air departs downwardly through the spaces between the blow boxes and then flows horizontally in the cross machine direction of the machine while absorbing some moisture from the upper side of the next underlying run of the web. At opposite sides of the machine the used drying air is sucked out by fans and sent through heating batteries and then recirculated to the blow boxes for renewed use.

Wet air is bled off the circulation, and dry air is admixed in order to maintain a sufficiently low moisture content of the circulating air. The temperature of the air ejected from the blow boxes is preferably held between about 100° to about 120°C and the wet temperature thereof may be about 60°C. By controlling these temperatures the total amount of evaporated moisture, or the drying capacity of section B_2, can be controlled and brought to suit varying working conditions.

In a papermaking machine with a designed web speed of about 1,500 fpm, for a 200-inch wide web, there are required about 600 blow boxes in the air drying section, which correspond to a freely suspended web length of about 450 feet (each blow box is about 6 inches in size and has a spacing from the next blow box of 4 inches). The arrangement of the blow boxes in five superimposed tiers reduces the length of section B_2 to about 90 feet, and its height may be about 4 feet.

In addition to the shrinkage in the cross machine as well as the machine direction of the web caused by drying while airborne, the paper web is also subjected to a mechanical shrinkage in its machine direction while having a moisture content preferably within the range of 28 to 42%, and most preferably about 30 to 40% moisture content. This mechanical shrinking is performed by a unit shown in the left part of Figure 8.4b, comprising a pair of rolls **21** and **23**. Roll **21** is heated and preferably has a steel surface, and roll **23** is rubber covered. The rolls are pressed together and driven at differential speeds, the rubber roll running slower.

A moist paper web passing the pressure nip of rolls **21, 23** will be compacted in its machine direction to an extent depending upon the line pressure of the nip and the speed differential of the rolls, whereas the width of the paper web is left substantially unaffected.

A paper web having a moisture content of between 28 and 42% can be mechanically shrunk in this unit, although the best results are obtained when the moisture is about 30 to 40%. With five runs of the paper web back and forth through the air drying section for a designed total drop in moisture content from 50 to 15%, a suitable moisture content of about 35% is obtained after the second run **12**, and therefore the web need not turn about the roll **17**, which may be removed, but is drawn through the nip of the compacting unit **21, 23** and then returned to the air drying section in order to make its third run therethrough.

The turning rolls **16, 17** and **19** are idling and essentially do not change the tension present in the web passing thereover. The turning roll **18** at the end of the third run of the paper web may cooperate with a driven roll **25** or even a nip with one driven roll, and the web may be drawn out of the section B_2 by a driven roll **27**. By controlling the speed of the rolls **25, 27** in accord with the speed of the compacting unit a desired tension can be maintained in the web, whereby the shrinkage of the web in the machine direction caused by the drying is partly or wholly eliminated and also a part of the mechanical shrinkage performed by the shrinking unit **21, 23** can be removed.

In order to prevent wrinkles and to straighten out the paper web in the cross machine direction at least some of the turning rolls may be replaced by bent rolls (Mont-Hope or Varibow). Such a roll **29** is shown close to the turning roll **19**, which it may replace. Another bent roll **31** is placed immediately before the compacting unit. These bent rolls may also effect a certain stretching of the web in the cross machine direction, whereby part of the transverse shrinkage caused by the drying of the web while airborne might be eliminated.

Example: A paper web was marked in a given area to produce a paper piece measuring 100 units long and 100 units wide, when leaving the former and web press A. During runs **11** and **12** through air drying section B_2, the paper web was free to shrink in the cross machine direction and was subjected to a low draw tension whereby the web was not too restricted in the machine direction. Just before entering the nip in the compacting unit **21, 23**, the marked paper web area measured 98 units long and 97 units wide. The compacting unit heating nip pressure was about 85 lb per linear inch of the nip and the speed differential of rolls **21, 23** was of about 3%, with the rubber coated roll **23** being the slower roll.

With these conditions the web was subjected to a compaction in the machine direction of about 3%, causing the marked paper web area to be about 95 units by 97 units when leaving the compaction unit. The roll **25** was driven with a speed which was 2% higher than the speed of the paper web leaving the compactor, and therefore in the third run (**13**) of the paper web through the air drying section B_2 the tendency of the web to shrink in the machine direction because of the drying or moisture removal was prevented. In fact, a greater part (about ⅔) of the shrinkage previously affecting the web in the compactor was eliminated.

Because of the draw tension thus maintained on the paper web the shrinkage of the web in the cross machine direction actually exceeded the value which would be obtained when no such draw tension was present. Because of this, when passing roll **18** the previously-mentioned marked paper web area was 97 units by 94 units. Roll **27** was driven with a somewhat lower speed than roll **25** (about 5% lower) with the paper being allowed to shrink naturally in the fourth and fifth runs (**14** and **15**), even in the machine direction. However, a sufficient tension was maintained to hold the paper web flattened out and to allow proper guidance of the web. The marked paper web area reached the air dry section B_2 with a length of 92 units and a width of 92 units and maintained these measurements during the subsequent final drying operation.

The paper thus obtained had an extensibility of about 8% in either of two directions at right angles with each other and therefore was highly useful for the production of paper bags or sacks. The paper was also made comparatively stiff as opposed to the paper produced by prior art process, and this stiffness favored the handling of the paper in bag-making machines.

Examples 2 through 9 are comparisons with prior art processes. Examples 2 through 5 relate to sulfate paper produced using four different processes and Examples 6 through 9 refer to sulfite paper manufactured in the same four different processes.

Example 2: A sulfate paper was manufactured in the conventional way without mechanical shrinkage and in a papermaking machine merely having cylinder dryers.

Example 3: A sulfate paper was manufactured from the same pulp as in Example 2 without mechanical shrinkage and in a papermaking machine having a drying section in which the paper web was airborne.

Example 4: A sulfate paper was manufactured from the same pulp as in Example 2 and in the same machine as there, but with a shrinkage unit inserted in the early part of the cylinder drying section of the machine and comprising a pair of cooperating rolls, one roll being rubber covered and driven at a 6 to 8% lower speed than the other hard-surfaced roll, whereby about 6 to 8% mechanical shrinkage was imparted to the paper.

Example 5: A sulfate paper was manufactured from the same pulp as in Example 2 in a paper machine of the kind previously described with reference to Figure 8.5 and comprising not only a shrinkage unit but also a drying section in which the web was airborne. The paper was mechanically shrunk 6 to 8%, in the machine direction. The drying section corresponded to that illustrated in Figure 8.4b, with the paper web dried to 35% moisture content entering the mechanical shrinking unit.

Examples 6 through 9: A sulfite paper was treated as according to the above Examples 2 through 5. However, the paper had a somewhat higher area weight, 90 grams per square meter, as compared with 70 grams per square meter of the sulfate paper. The process of treating paper in Examples 6 through 9 corresponded to Examples 2 through 5, respectively.

All the papers obtained above were tested for tensile strength, ultimate elongation and fracture energy. The results obtained for both machine and cross machine direction are reported in the table on the following page.

Example	Tensile Strength (kilogram-force)		Ultimate Elongation (percent)		Fracture Energy (kilogram-force/cm)	
	Longitudinal*	Transverse**	Longitudinal	Transverse	Longitudinal	Transverse
2	9.2	4.8	3.5	5.0	1.0	0.8
3	9.0	4.1	7.6	11.1	1.8	1.4
4	5.9	4.8	8.5	6.3	2.9	1.1
5	5.8	4.2	8.1	9.7	2.7	1.3
6	11.1	5.0	1.7	4.3	1.2	1.6
7	8.3	4.7	2.3	6.6	1.2	2.0
8	10.0	4.3	3.3	3.7	1.9	1.3
9	9.7	3.9	7.8	6.3	3.3	1.7

*Machine. **Cross machine.

The above examples show the improvement of the paper qualities made possible by the use of the process.

OTHER PAPERS

ERASABLE PAPERS

Acid Treatment plus Perfluorinated Carboxylic Acid Surface Size

W.R. Wasko, M.R. Sheppard and R.J. Eber; U.S. Patent 3,562,099; February 9, 1971 found that, by subjecting paper to contact for a defined period of time with a relatively concentrated aqueous sulfuric acid solution of defined strength, the paper is rendered more readily erasable, particularly with respect to typewriting.

Paper base stocks, or "furnishes" which have been modified by the addition of certain additives provide superior and therefore preferred base stocks for making more readily erasable paper.

The process results in the production of papers which exhibit superior whiteness and opacity, as well as superior erasability, rendering them particularly suitable for commercial sale and use as a writing paper and particularly as a typewriter paper.

Furthermore, it also has been found that through the application of a small amount of one or more selected sizing agents to the surfaces of the paper after acid contact and removal, and after at least partial drying, the tendency of typing inks to "feather" or "wick" on the product paper is markedly reduced over that observed with such paper which is not so coated.

The improved paper stock is prepared by incorporating one, some or preferably all of the following additive materials in a stock comprising rag or wood pulp: (1) a small amount, sufficient to reduce significantly paper transparency, of an opacifying filler or pigment; (2) a small amount, sufficient to reduce significantly paper absorbency, of a sizing agent; (3) a small amount, sufficient to improve significantly paper wet strength, of a wet strength improving agent; and (4) a small amount, sufficient to improve optical brightness, of an optical brightening agent.

The additive materials are incorporated in the stock preparation stage, i.e., during the conventional beating or refining operations, or in both. In this manner the additives are uniformly distributed throughout the stock, and ultimately throughout the paper after the latter is formed by conventional techniques and machines.

Contacting between the concentrated aqueous solution of sulfuric acid and the paper can be effected by suitable conventional techniques and apparatus. After the paper has been sufficiently contacted with the concentrated aqueous solution of sulfuric acid, it can be subjected to conventional techniques and apparatus to remove substantially all of the remaining acid solution.

The acid contacting in accordance with the process of a paper made from either a nonmodified or modified paper stock produces a marked improvement in the erasability of such paper, particularly with respect to typewriting. The modification of the paper stock provides an additional benefit in that the product paper exhibits superior opacity, superior whiteness or brightness, and greater wet strength, all of which makes such paper more attractive from a commercial standpoint.

However, it also has been observed that a paper rendered more readily erasable by practice of only the acid-contacting technique still exhibits an undesirable susceptibility to smudging by the usual typing inks, and to wicking or feathering by such inks, including after erasure and retyping.

It also has been found that at least the tendency of the acid-treated paper to promote ink wicking or feathering can be substantially eliminated or significantly reduced by the technique of applying one or more selected sizing agents to the surface of the acid contacted paper following acid removal and washing, and after at least partial drying of the paper.

The suitable sizing agents are the fluorinated carboxylic acids, including the perfluorinated carboxylic acids, used either alone or in combination with a suitable starch or water-soluble gum. Suitable starches include unmodified starches and modified starches such as chlorinated starches, ethylated starches, acetylated starches.

Suitable water-soluble gums include such substances as carboxymethylcellulose, methylcellulose, ethylcellulose, polyvinyl alcohol, latex emulsions, animal glues, proteins such as casein and soy protein. Minnesota Mining & Manufacturing Company's "FC 806" has been found to be particularly satisfactory and is preferred for use. (This product consists essentially of a fluorocarbon material.)

The surface application of the sizing agent can be effected by conventional techniques and sizing equipment. Generally, the sizing agent is applied in the form of a solution, typically aqueous. In the case of FC 806, for example, such solution consists of from about 0.1 to 0.5% (preferably 0.25%) by weight of the active agent of FC 806 in water.

The aqueous solution of the sizing agent is applied to the at least partially-dried web of acid contacted paper through the nip of a conventional size press applicator and in an amount sufficient to result in the addition on a water-free basis of from about 0.01 to 0.05 pound of the fluorocarbon sizing agent for each 3,000 square feet of paper (i.e., 500 sheets, each 24" x 36").

In the case of FC 806, the addition of about 0.025 pound of the active sizing agent in that product on a water-free basis per 3,000 square feet has been found to give excellent results.

Although a fluorocarbon sizing agent such as FC 806 produces excellent results when employed alone, it also may be used together with other materials such as suitable starches or water-soluble gums. When such an additional material is used, it also is applied in the aqueous solution and therein at a concentration of from about 2 to 10% by weight, with 6% being typical. At such concentrations, the additional material is applied to the paper web at a rate equivalent, on a dry basis, to from 0.2 to 1.0, and typically 0.5, pound of the additional material per 3,000 square feet of paper stock.

Example: (a) Modified Furnish — A paper base stock or furnish is prepared in a conventional manner comprising 100 weight percent of a bleached hardwood pulp available under the name Espanola. This pulp is suitable for manufacturing an absorbent type paper (commonly refered to as water leaf), which typically is used in the manufacture of parchment paper.

This furnish is modified so as to produce a superior base stock for making a more readily erasible paper as herein disclosed. Specifically, there is incorporated in such furnish about 5% of a titanium dioxide opacifying filler available commercially under the name Rayox, about 0.25% of a rosin size available under the name Cyfor, about 0.25% of a melamine-formaldehyde resin wet strength agent available under the name Parez 607, and about 0.0625% of an optical brightening agent available under the name Leucophor BS.

The aforesaid percentages are by weight, and are based on dry pulp. After suitable preparation steps, such as beating, refining or both, the modified furnish so prepared is formed into an absorbent paper sheet material by conventional machines and techniques of the Fourdrinier type.

(b) Acid Treatment — The paper sheet material formed as above described is continuously contacted with a relatively concentrated aqueous solution of sulfuric acid containing about 68% by weight (range 65 to 72%) of sulfuric acid, and is allowed to remain in contact therewith for a total contact time of about 5 seconds. This contact is effected by continuously running the paper sheet material or web, submerged, through a liquid phase or bath consisting essentially of the above-specified, relatively concentrated aqueous sulfuric acid solution, and, after emergence, allowing the paper to remain substantially wet with the acid solution for the balance of the specified 5-second contact time. The temperature of the bath, i.e., the contact temperature, is maintained relatively constant at about 60°F (range 55° to 65°F).

After emergence from the relatively concentrated aqueous sulfuric acid solution, the paper is continuously contacted with a relatively and substantially more dilute aqueous solution of sulfuric acid, typically containing approximately 35% or less by weight of sulfuric acid. As before, contacting conveniently can be effected by continuously passing the paper sheet material, submerged, through a liquid bath consisting essentially of the relatively dilute aqueous sulfuric acid.

However, this contacting preferably is effected countercurrently, by passing the paper continuously through a countercurrent liquid stream which consists initially of water. In this case of contacting with dilute acid solution, the contacting need only be sufficiently long to accomplish an adequate mixing of the two acid solutions. The contact temperature is maintained in the range of from 50° to 100°F.

After the paper sheet material emerges from the relatively dilute aqueous sulfuric acid solution, it is thoroughly washed, if necessary with an excess of water, until substantially all of the acid is removed. Prior to a final water wash the paper web is contacted with an aqueous ammonium hydroxide solution to remove remaining traces of acid and adjust the pH of the paper to about 8.0 to 8.5.

The resulting paper sheet material, substantially free of residual acid, then is subjected to drying by means of conventional equipment and techniques.

(c) Surface Sizing – After at least partial drying, the surface application of a sizing agent is effected in the nip of a conventional size press applicator, through which the at least partially dried web of acid-treated paper is progressing. The sizing agent employed is the fluorocarbon product available from the Minnesota Mining & Manufacturing Company under the name FC 806.

The material for application consists essentially of an aqueous solution containing about 0.25% by weight of the active sizing agent of FC 806 in water. This aqueous solution of sizing agent is applied at the rate on a dry basis, of about 0.025 pound of solute fluorocarbon per 3,000 square feet of paper. After application of the sizing agent, the web is further dried by means of conventional equipment and techniques.

(d) Evaluation – The paper prepared as above described then is subjected to comparative testing: (1) with paper prepared identically and from the same paper stock which paper has been acid contacted but has not been given a surface application of the sizing agent; and (2) with a conventional and standard 25% rag pulp bond paper available commercially and illustrative of a typical high quality typewriter paper.

To each of such specimens identical typewriting then is applied in an identical manner and utilizing a single-use, carbon ribbon. Each specimen then is examined or tested in substantially identical manner for, among other things, erasability and resistance to ink feathering or wicking. The results of all such tests are evaluated and compared visually and in the same manner. The following tabulation illustrates the results which are achieved.

Paper Stock	Erasability	Resistance to Feathering
Conventional (untreated) bond paper	Poor	Excellent
Acid-treated paper	Good	Poor
Acid-treated and surface-sized paper	Excellent	Excellent

As will be seen from the foregoing results, the product paper produced by practice of the acid contacting technique exhibits in both instances an erasability which is superior to that possessed by the conventional bond typewriter paper which is not so treated.

However, it also will be noted that the paper stock which is subjected only to the acid contacting step is inferior to the untreated conventional bond paper with respect to its resistance to ink feathering. Nevertheless, the paper stock which is subjected to the practice of the process in its preferred form, i.e., a combination of acid treatment and surface sizing, exhibits excellent resistance to ink feathering in addition to superior erasability.

The observed excellence in resistance to ink feathering is particularly desirable, and unexpected, in view of the somewhat poorer results in this regard achieved with other sizing agents tested. Results substantially similar in all respects are achieved when the same tests are conducted utilizing a cloth multi-use, typewriter ribbon instead of a carbon ribbon.

It also should be noted that each of the papers described above, being made from a furnish which had been modified with additives in accordance with the process exhibits an opacity and whiteness rendering it of relatively superior quality for commercial sale and use as a typewriter paper.

Acid Treatment plus Alkanolamine Salt of Fluorine-Substituted Aliphatic Phosphate Surface Size

In this process of *W.R. Wasko and M.R. Sheppard; U.S. Patent 3,562,100; February 9, 1971* the suitable surface sizing agents are the alkanolamine salts of fluorine-substituted aliphatic phosphates, including primary, secondary and tertiary alkanolamine salts of mono- and poly-fluoroalkyl phosphates.

A specific sizing agent of the fluorochemical type found to be particularly preferable is mixed diethanolamine salts of mono- and bis(1H,1H,2H,2H-perfluoroalkyl) phosphates where the alkyl group contains an even number of carbon atoms in the range of from 8 to 18 and the salts have a fluorine content of from 52.4 to 54.4%, as determined on a solids basis. This product is available as Zonyl RP.

The fluorochemical sizing agent can be employed alone, or together with a suitable unmodified starch, a modified starch other than a chlorinated starch, a water-soluble gum, or a mixture of any of the foregoing.

The surface application of the sizing agent can be effected conveniently by conventional techniques and sizing equipment. Generally, the sizing agent is applied in the form of a solution, typically aqueous. In the case of Zonyl RP, for example, such solution consists of from about 0.1 to 0.5% (preferably 0.25%) by weight, of the active agent of Zonyl RP in water.

The aqueous solution of the sizing agent is applied to the at least partially-dried web of acid-contacted paper through the nip of a conventional size press applicator, and in an amount sufficient to result in the addition on a water-free basis of from about 0.01 to 0.05 pound of the fluorochemical sizing agent for each 3,000 square feet of paper (i.e., 500 sheets, each 24" x 36"). In the case of Zonyl RP, the addition of about 0.025 pound of the active sizing agent in that product on a water-free basis per 3,000 square feet has been found to give excellent results.

Although a fluorochemical sizing agent such as Zonyl RP produces excellent results when employed alone, it also may be used together with other materials such as suitable starches or water-soluble gums. When such an additional material is used, it also is applied in the aqueous solution and therein at a concentration of from 2 to 10% by weight, with 6% being typical. At such concentrations, the additional material is applied to the paper web at a rate equivalent, on a dry basis, to from about 0.2 to about 1.0, and typically 0.5, pound of the additional material per 3,000 square feet of paper stock.

Example: (a) Modified Furnish and (b) Acid Treatment have been described in U.S. Patent 3,562,099 above.

(c) Surface Sizing – After at least partial drying, the surface application of a sizing agent is effected in the nip of a conventional size press applicator, through which the at least partially dried web of acid-treated paper is progressing. The sizing agent employed is the fluorochemical product available under the name Zonyl RP.

The material for application consists essentially of an aqueous solution containing approximately 0.25% by weight of the active sizing agent of Zonyl RP in water. This aqueous solution of sizing agent is applied at the rate, on a dry basis, of approximately 0.025 pound of solute fluorochemical per 3,000 square feet of paper. After application of the sizing agent, the web is further dried by means of conventional equipment and techniques.

(d) Evaluation – as described above in U.S. Patent 3,562,099. The following tabulation illustrates the results which are achieved.

Paper Stock	Erasability	Resistance to Smudging	Resistance to Feathering
Conventional (untreated) bond paper	Poor	Good	Excellent
Acid-treated paper	Good	Poor	Poor
Acid-treated and surface-sized paper	Excellent	Good	Excellent

As will be seen from the foregoing results, the product paper produced by practice of the acid-contacting technique exhibits in both instances an erasability which is superior to that possessed by the conventional bond typewriter paper which is not so treated.

However, it also will be noted that the paper stock which is subjected only to the acid contacting step is inferior to the untreated conventional bond paper with respect to its resistance to ink smudging and feathering. Nevertheless, the paper stock which is subjected to the practice of the process in its preferred form, i.e., a combination of acid treatment and surface sizing, exhibits good resistance to ink smudging and excellent resistance to ink feathering in addition to superior erasability. Results in most instances substantially similar to the above are achieved when the same tests are conducted utilizing a cloth, multi-use, typewriter ribbon instead of a carbon ribbon.

It also should be noted that each of the papers described above, being made from a furnish which had been modified with additives in accordance with the process, exhibits an opacity and whiteness rendering it of relatively superior quality for commercial sale and use as a typewriter paper.

Acid Treatment plus Low Molecular Weight Polyester Surface Size

The surface sizing agents here found most suitable by *W.R. Wasko, M.R. Sheppard, O.R. Schultz and S.A. Albertson; U.S. Patent 3,607,621; September 21, 1971; assigned to Paterson Parchment Paper Company* are the relatively low molecular weight polyesters made by reaction of rosin and maleic anhydride adducts with polyhydric alcohols or polyols and including, for example, such polyesters with aliphatic diols, triols and other polyols such as ethylene glycol, glycerin, sorbitol, mannitol, pentaerythritol and the like, with aromatic polyols such as phenol, catechol, resorcinol and hydroquinol and with mixtures thereof.

Specific sizing agents found particularly suitable are polyesters made by reaction of rosin-maleic anhydride adducts with aliphatic polyols such as ethylene glycol and glycerin, or with a mixture of two or more such polyols. Suitable agents of this type are well known, and many are commercially available. For example, the product sold under the name Shanco 4100 Resin has been found to be particularly satisfactory and is preferred for use. This material consists essentially of polyesters of the type herein disclosed as suitable for use as a sizing agent, and it has the following typical properties:

Molecular weight	1,000-2,000
Melting point (Ring and Ball Method), °C	147-157
Acid number	183-193
Color, Gardner (50% in toluol)	5-7
Specific gravity 25°C	1.14

The surface application of the sizing agent can be effected conveniently by conventional techniques and conventional equipment such as sizing equipment. Generally, the sizing agent is applied in the form of a solution, typically aqueous and alkaline. In the case of Shanco 4100 Resin such solution contains 4.0 to 16.0% by weight of the resin in a dilute aqueous alkali (pH 8 to 8.5), such as aqueous ammonium hydroxide; a concentration of 8.0% is preferred.

The aqueous alkaline solution of the sizing agent is applied to the at least partially dried web of acid contacted paper through the nip of a conventional size press applicator, and in an amount sufficient to result in the addition on a water-free basis of from about 0.5 to about 2.0 pounds of the sizing agent for each 3,000 square feet of paper (i.e., 500 sheets, each 24" x 36"), In the case of Shanco 4100 Resin, the addition of about 1.0 pound of that sizing agent on a water-free basis per 3,000 square feet has been found to give excellent results. Although a sizing agent such as Shanco 4100 Resin produces excellent results when employed alone, it also may be used in combination with other suitable sizing agents.

Example: (a) Modified Furnish and (b) Acid Treatment have been described in U.S. Patent 3,562,099.

(c) Surface Sizing – After at least partial drying, the surface application of a sizing agent is effected in the nip of a conventional-size press applicator, through which the at least partially dried web of acid-treated paper is progressing. The sizing agent employed is 8.0 wt % of a product available commercially under the name of Shanco 4100 Resin in dilute aqueous ammonium hydroxide (pH of

8.0 to 8.5). This solution of sizing agent is applied at the rate, on a dry basis, of approximately 1.0 pound of the solute resin per 3,000 square feet of paper. After application of the sizing agent, the web is further dried by means of conventional equipment and techniques.

(d) Evaluation – The paper prepared as above described then is subjected to comparative testing: (1) with paper prepared identically and from the same paper stock, which paper has been acid contacted but has not been given a surface application of the sizing agent; and (2) with a conventional and standard 25% rag pulp bond paper available commercially and illustrative of a typical high quality typewriting paper.

To each of such specimens identical typewriting then is applied in an identical manner and utilizing a single-use, carbon ribbon. Each specimen then is examined or tested in substantially identical manner for (1) uniformity or evenness of typewriting; (2) erasability; (3) resistance to ink smudging; and (4) resistance to ink feathering or wicking. The results of all such tests are evaluated and compared visually and in the same manner. The following tabulation illustrates the results which are achieved:

Paper Stock	Erasability	Resistance to Smudging	Resistance to Feathering	Uniformity of Typewriting
Conventional (untreated) bond paper	Poor	Good	Excellent	Good
Acid-treated paper	Good	Poor	Poor	Good
Acid-treated and surface-sized paper	Good	Excellent	Excellent	Excellent

As will be seen from the foregoing results, the product paper produced by practice of the acid contacting technique of the process exhibits in both instances an erasability which is superior to that possessed by the conventional bond typewriter paper which is not so treated.

However, it also will be noted that the paper stock which is subjected only to the acid contacting step is inferior to the untreated conventional bond paper with respect to its resistance to ink smudging and feathering. Nevertheless, the paper stock which is subjected to the practice of the process in its preferred form, i.e., a combination of acid treatment and surface sizing, exhibits excellent resistance to both ink smudging and feathering, and excellent receptivity to typewriting, in addition to superior erasability.

The observed superiority in evenness of type is particularly desirable, and unexpected, in view of the somewhat poorer results in this regard achieved with other sizing agents tested. Results substantially similar in all respects are achieved when the same tests are conducted utilizing a cloth, multi-use, typewriter ribbon instead of a carbon ribbon.

It also should be noted that each of the papers described above, being made from a furnish which had been modified with additives in accordance with the process, exhibits an opacity and whiteness rendering it of relatively superior quality for commercial sale and use as a typewriter paper.

Use of 60 to 97% Hydrated Gelatinized Cellulosic Fibers with 3 to 40% Unhydrated Fibers

The process of *W.G. Louden; U.S. Patent 3,839,144; October 1, 1974* relates to a paper sheet consisting essentially of a pressed and dried water-laid web of cellulosic papermaking fibers where the fibers in the web are a blend of substantially unhydrated cellulosic papermaking fibers and of hydrated, gelatinized cellulosic papermaking fibers.

There is present in the furnish at the time the web is formed substantially unhydrated cellulosic papermaking fibers blended with the hydrated, gelatinized fibers in a proportion of from about 3, preferably from about 5, to about 40% by weight, of the substantially unhydrated fibers to from about 97, preferably from about 95, to about 60% of the hydrated, gelatinized fibers.

The provision of the stated blend of fibers may be accomplished in either of two general procedures; (1) subjecting all fibers to a common refining operation, the fibers which are to be hydrated and gelatinized being of the type that are relatively quick and easy to hydrate and the fibers which are to be relatively unhydrated being of the type which are relatively difficult and slow to hydrate; or (2) subjecting only the fibers which are to be hydrated and gelatinized to refining and then mixing relatively unhydrated fibers therewith just prior to sheet formation. Conventional refining apparatus used for dense papers may be used in method (1).

One type of fiber subjected to refining will be of a type that is relatively quick and easy to hydrate and gelatinize, namely wood pulp containing less than 92%, preferably in the range of from 85 to 89%, by weight of alpha-cellulose, the balance being essentially hemicelluloses.

Examples of such pulps are sulfate (kraft) pulps, Mitscherlich sulfite pulps, semichemical pulps and soda pulps having the stated alpha-cellulose and hemicellulose contents. The other type of fiber which is present during refining is of a type which is slow and difficult to hydrate, namely high alpha-cellulose fibers (at least 92% alpha-cellulose) such as cotton fibers (from linters, new cloth or old rags), flax fibers, hemp fibers, rayon (regenerated cellulose) fibers, wood pulps having an alpha-cellulose content of at least 92% and the like. Preferably the relatively unhydrated fibers will be long papermaking fibers. The fibers used for both types may be bleached or unbleached. Cotton fibers are the preferred high alpha-cellulose fibers.

In the second method referred to above, that is, where the relatively unhydrated fibers are mixed with the hydrated, gelatinized fibers just prior to sheet formation, the hydrated, gelatinized fibers will be of the same type as described above.

The relatively unhydrated fibers may be any type of cellulosic papermaking fibers, including each of the types mentioned above in connection with the first method, as long as they have not been refined if they are of the type that is easily and quickly hydrated and gelatinized.

Hydration as used herein refers to the enhancement of the fiber's affinity for water brought about by beating or refining. This procedure also causes swelling (gelatinization), cutting, fraying and fibrillation. One way of distinguishing between the two types of fibers is on the basis of differences in freeness caused by variation in the degree of hydration. A substantially unhydrated pulp will allow water to drain therefrom relatively quickly, whereas a well beaten pulp (hydrated) will lose water slowly.

The hydrated, gelatinized fibers used will be hydrated to the extent equivalent to a Schopper-Riegler freeness of less than 340 cc, and preferably from about 180 to about 280 cc, that is substantially to the extent of fiber used in the manufacture of grease-proof paper. The substantially unhydrated fibers, on the other hand, will be equivalent to a Schopper-Riegler freeness of at least 680 cc, and preferably from about 800 to about 850 cc.

Examples 1 through 4: A bleached northern kraft pulp of about 89% alpha-cellulose content was hydrated to a Schopper-Riegler freeness of about 200 cc in a continuous Jordan-type refiner as used in the manufacture of grease-proof paper (fiber A).

Separately, a bleached kraft pulp of 97.5% alpha-cellulose content and having a Schopper-Riegler freeness of about 800 cc was dispersed in water (fiber B). Portions of each of these fiber dispersions were blended together, in proportions as set forth in the following table, and hand sheets were made according to TAPPI Standard Procedure T-205-m-58. Test results are set forth below (bearing in mind that hand sheets are normally less dense than machine-made paper using the same papermaking fibers).

Example	1	2	3	4
Percent fiber A	100	90	80	70
Percent fiber B	0	10	20	30
Caliper (inches)	0.0032	0.0034	0.0039	0.0040
Opacity (TAPPI T-425-OS-60)	60	63	74	70
Density (g./cc.)	0.74	0.69	0.62	0.59
Erasability	excellent	excellent	fair*	poor*

*Paper made from these fibers on a paper-making machine would be expected to have better erasability than these hand sheets.

Similar results can be observed by employing, in place of the 97.5% alpha bleached northern kraft pulp, cotton or rag fibers. Likewise, unrefined regular kraft or sulfite pulps (85 to 92% alpha-cellulose content, and a Schopper-Riegler freeness above 680 cc) and unrefined semichemical or soda pulps (Schopper-Riegler freeness above 680 cc) can also be used in place of the 97.5% alpha bleached northern kraft pulp, although the effect on caliper and density is a little less pronounced.

Erasability characteristics are determined by typing an image on the paper; erasing it with a lead pencil eraser, noting the image remaining and damage to paper, if any; retyping over the same area and noting the residue of the original image, if any. Results are characterized as follows: excellent—faint trace of original image and no damage to paper after erasing but original image not discernible after retyping; good—same as excellent but faint evidence of original image after retyping; fair—same as good but original image fairly obvious after retyping; and poor—sheet damaged on erasing and original image is pronounced after retyping. Conventional typewriter letter head papers would have an erasability rating under this criterion of poor. Specifications for typical paper grades or uses producible according to the process are as follows:

	Proportion (%) of unhydrated fibers		base wt. (lbs.)
	general	preferred	(17 × 22", 500)
erasable bond	5 - 40	10 - 25	6 - 30
tracing paper	5 - 40	5 - 15	6 - 24
master sheets for diazo printing	5 - 40	5 - 15	6 - 24
printing paper	5 - 40	10 - 25	6 - 50
cover stock	5 - 40	15 - 35	50 - 100
base paper for release coating	5 - 40	15 - 35	6 - 50

ANTISTATIC AND CODE PAPERS

Inorganic Powder Impregnated with Antistatic Surface Agent of HLB Below 9

T. Yasuda; U.S. Patent 3,682,696; August 8, 1972; assigned to Tatsuta Electric Wire & Cable Company Limited, Japan is concerned with preventing paper from being charged with static electricity without exerting any harmful influence upon the physical and chemical characteristics thereof.

In the method, a small quantity of a powder of a nonelectrifiable inorganic substance, obtained by dissolving a surface active agent of HLB (hydrophile-lypophile balance) below 9 in an organic solvent and then causing it to adhere to the surface of the powder, is added either to the pulp in the paper manufacturing process or the coating material in the paper coating process, so that the powder will adhere to the inside or the surface of the paper uniformly.

The reason why the HLB of the electrification preventive surface active agent is restricted to less than 9 is that if the HLB of the agent is very high, it is liable to be separated from the inorganic substance powder by the water which is used in large quantities in the paper manufacturing process or the coating material preparing process, with the result that the effect of the agent will be lost.

The antistatic surface active agent comprises at least one member selected from the group consisting of aliphatic amine salts, aliphatic ammonium salts, polyethyleneglycol esters, polyethyleneglycol ethers, alkylphenol/polyethyleneglycol condensates, polyethyleneglycol alkylamides, polyethyleneglycol alkylamines, polyhydric alcohol esters, ester/ether condensates, or alkylbetaines.

The inorganic powder is selected from the group consisting of active alumina, kaolin, clay, talc, aluminum hydroxide, gypsum, titanium oxide, magnesium oxide, calcium carbonate, magnesium carbonate, zinc sulfide, silica, calcium silicate, magnesium silicate, satin white, and diatomaceous earth.

Example 1: 2.5 grams of a polyethyleneglycol alkylamine whose HLB is below 9 was added to 1,000 cc of ethyl alcohol and completely dissolved by heating at approximately 60°C. 1,000 grams of dry active alumina powder with an average diameter of 1 micron was added to the solution.

The resultant mixture was agitated and then placed in a constant temperature oven of approximately 80°C so as to remove ethyl alcohol by evaporation. By the foregoing process there was obtained alumina powder on which was adsorbed the electrification preventitive surface active agent whose HLB is less than 9. Twenty parts of this powder was added to 100 parts of BSP at 60°SR in the beater of paper manufacturing process to obtain antistatic paper.

Example 2: Art Paper – 100 parts of the alumina powder obtained in Example 1, on which had been adsorbed the electrification preventive surface active agent of HLB less than 9 was added to 100 parts of clay coating material prepared with starch as a binder. The coating material was so coated that the coating quantity was 15 parts as against 100 parts in dry weight of BSP at 60°SR. Antistatic art paper was obtained.

Antistatic Paper for Punch Coding Purposes

M.L. DeMatte; U.S. Patent 3,902,959; September 2, 1975; assigned to Westvaco Corporation found that the addition of a small amount of an inorganic salt, in a surface size press treatment to code paper during the papermaking process, will eliminate altogether the buildup of static electricity charges on the tapes prepared from the code paper during the punching and perforating operation.

This is achieved without leaving a residual grit in the paper or otherwise producing any other deleterious effect on the physical properties of the paper including its lubricity where the inorganic salt is incorporated in the production of code paper containing oil.

A dispersion of starch in water is first prepared. Then under agitation and in the presence of heat, an inorganic salt is added to the dispersion. The dispersion thus produced is subsequently applied to the base paper web by any suitable means such as dipping, spraying or brushing, but preferably is applied by means of a size press or a papermachine to produce a reliable antistatic code paper.

In the case where the code paper is desired to have improved lubricity, an emulsifier and a lubricating oil are also added to the dispersion to form an oil in water emulsion substantially as disclosed in U.S. Patent 3,702,801. After the paper is impregnated with either the starch/salt dispersion or the oil-in-water/starch/salt dispersion at the size press, it is dried to a moisture content of about 4 to 8% and slit into the required dimensions for code paper tapes.

Pearl, acid modified, cationic, oxidized and ethylated starches can be used in the sizing, with the preferred being a pearl corn starch. The amount of starch in the size press formulation can be varied from about 1.5 to 15% by weight, with the preferred being about 5 to 10% by weight.

Where a prelubricated antistatic code paper is produced, the oil used may be one of any of the well-known liquid lubricating oils, including motor oils, paraffin based oils, and mineral oils. The amount of oil used in the emulsion is quite small, ranging from about 1 to 10%, and preferably from about 3 to 5% by weight of the total emulsion including starch, water, emulsifier and oil. The emulsifiers used in making the prelubricated antistatic code paper may be of any of those commercially available and particularly an emulsifier known as Tween 80, a polyoxyethylene sorbitan monooleate or Tegitol-NP27, a nonylphenylethylene glycol ether. Acid salts, basic salts and neutral salts would each perform satisfactorily but salts from the group consisting of ammonium chloride, potassium chloride and sodium chloride are preferred.

Concentrations as low as 0.3% by weight of salt when applied at the size press under mill conditions, produced paper having very good antistatic properties without inhibiting the lubricity or other physical properties of the paper. 0.3 to 6% by weight of salt is added. The chemical wood pulp used is preferably bleached to produce a white pulp.

Example 1: A production trial of code paper was conducted at the Westvaco Corporation Paper Mill in Tyrone, Pennsylvania (No. 4 papermachine). Three different conditions were evaluated. In the first case, an 8.3% by weight Pearl X starch solution was used to produce control sheets. The pickup from this formulation was about 1.5 lb/ream.

A second condition comprised 8.3% Pearl X starch and 0.6% potassium chloride (KCl) which produced a pickup of 1.6 lb/ream. The third trial condition comprised 8.3% Pearl X starch and 0.3% KCl which produced a pickup of about 1.6 lb/ream with all percentages being by weight.

The paper thus produced was then dried, cut into sheets and subjected to electrometer tracings at a relative humidity of 16% in order to determine the effectiveness of the antistatic agents applied. The results of the tests are reproduced below:

Condition	Charge Acceptance (volts)	Time to Decay (sec)	Burpee Punch (visual)
Control	300	14	Static
0.6% KCl	160-240	1.0-1.8	No static
0.3% KCl	240-360	2.4-4.6	No static

The results of the test data make it clear that for the production of code paper on the paper machine, suitable antistatic properties are readily achieved with the inorganic salt KCl in rather small quantities. Another test performed on paper samples prepared during the above trial was the Burpee Punch test. For this purpose, the paper samples were conditioned at 20% RH and punched with a Burpee Punch.

The Burpee Punch is commonly used to introduce and store information on code paper tapes by punching a sequence of holes in the paper tape. In the past, it was found that code paper which has accumulated a static charge presented problems when punched on the Burpee unit since the chads punched from the paper tended to accumulate and cling to one another. However, only the control sheets caused static problems.

Example 2: In yet another machine trial on No. 4 papermachine at Westvaco's Tyrone, Pennsylvania paper mill, two additional size press conditions were used in the manufacture of Westvaco's Intra-Lube Code Paper. For this example, the size press formulation fully described in U.S. Patent 3,702,801 was used. Thus the size press solutions for the first condition comprised 8.3% Pearl X starch, 4% Faxam 40 lubricating oil and 0.05% emulsifier Tergitol NP-27, all percentages by weight.

In the second condition, everything remained the same except that 0.6% KCl was added to the formulation. The paper made under the above noted conditions was then sheeted and the samples used to prepare electrometer tracings in order to determine the effectiveness of the different treatments.

In each case the amount of size press formulation pickup varied from 1.5 to about 1.8 lb/ream. In addition, code paper samples from the two conditions were also subjected to the Burpee Punch test and the visual results were noted. The data obtained from the test is set forth below:

Condition	Charge Acceptance (volts)	Time to Decay (sec)	Burpee Punch (visual)
Control	300	14.6	Static
0.6% KCl	200	2.0-2.2	No static

Lubricated Paper

The paper produced by the process developed by *M.L. DeMatte; U.S. Patent 3,702,801; November 14, 1972; assigned to Westvaco Corporation* is intended for use in any operation which requires that the paper be cut, punched or trimmed.

These finishing operations or end use functions of the paper are performed with knives, both rotating and fixed, or with dies and punches, operating either in a punch type die-press or in a rotary press. In each instance, the paper, or several blanks of the paper must be cut or punched smoothly and clearly without tearing or splitting the paper.

Furthermore, for efficient and economical operation, the paper must exhibit good release characteristics from the die or punch type presses, and, the antistatic properties of the paper must be superior to insure runnability. Finally, however, one of the most important factors in developing a paper suitable for the cutting, punching or trimming operations hereinbefore mentioned, is the ability of the paper to perform in such operations satisfactorily without causing undue wear on the element which effects the particular operation to the paper.

Based on actual production tests of the prelubricated paper produced by the process the following improvements were noted: (1) longer die wear, (2) higher lifts cut, (3) ease of cutting and release from the die, (4) substantially no top to bottom variation in cutting, (5) better quality cutting, i.e., elimination of pressure points, pronounced burr, and no chipping, cracking or splitting, (6) considerably fewer damaged or sprung dies, (7) a somewhat diminished noise level on hard to cut paper, and (8) better panel cutting on window forming machines.

A dispersion of starch in water is first prepared. Under agitation, an emulsifier is added to the aqueous starch dispersion and then a lubricating oil is slowly added to produce an oil-in-water emulsion. The resultant oil-in-water emulsion may be applied to the base paper web by any suitable means such as dipping, spraying, and brushing, but preferably is applied by means of a conventional or gateroll size press on a papermachine.

The paper is impregnated with the emulsion at the size press and is subsequently dried to a moisture content of about 4 to 8%. The resultant prelubricated paper is then suitable for use in any cutting, punching or trimming operation without the necessity of any further oil treatment of the paper. Furthermore, the oil is uniformly contained within and on the paper in an amount which heretofore could not be applied uniformly to paper in a separate oiling treatment.

Pearl, the acid modified, cationic, oxidized, and ethylated starches can be used in the sizing, with the preferred being pearl corn starch. The amount of starch in the oil-in-water emulsion can be varied from about 1.5% to 15% by weight, with the preferred being about 5% to 7% by weight.

The oil used can be any of the known liquid lubricating oils, including motor oils, paraffin-based oils, and mineral oils. The viscosity of the oil component can be quite variable and may for example range from a Brookfield of about 27 centipoises (No. 1 spindle, 20 rpm) to about 870 centipoises (No. 2 spindle, 20 rpm) when measured at about 25°C.

The amount of oil in the emulsion is quite small, ranging from about 1 to 10%, and preferably from about 3 to 5%, by weight of the total emulsion including starch, water, emulsifier and oil. At these concentrations, the amount of oil picked up by the paper at the size press has ranged from about 0.2 pound to 1.5 pounds per ream, with the preferred pickup being in the range of about 0.5 to 1.1 pounds per ream of paper (500 sheets, 25" x 38").

Obviously, the amount of oil used can be above that stated, but oil odor and oil bleeding may occur as well as losses in opacity when increased amounts of oil are used. Oil applied in accordance with the process is uniformly distributed throughout the paper even though it is present in a very limited quantity.

Any suitable emulsifier can be used in sufficient quantity to effect an oil-in-water emulsion. Preferred are such emulsifiers known commercially as Tween 80, a polyoxyethylene sorbitan monooleate, and Tergitol-NP27, a nonylphenylethylene glycol ether. The emulsifier preferably is added to the aqueous starch dispersion before the oil is added thereto. The chemical wood pulp used is preferably bleached to produce a white pulp.

Example: An aqueous dispersion of pearl starch was prepared by cooking the starch as an aqueous slurry of about 5 to 10% solids in a continuous jet cooker at a temperature of about 320°F, pressure of about 80 psi, for about 35 minutes, and then diluting the resultant cooked starch with sufficient water to produce 1,000 gallons of a 6% by weight aqueous dispersion of starch.

Under agitation, 1 gallon of Tergitol-NP27 emulsifing agent was added to the starch dispersion. Then with continuous agitation, 40 gallons of Faxam-40 lubricating oil (one of a line of lubricating oils produced by Humble Oil & Refining Company), having a Brookfield viscosity of about 40 centipoises (No. 1 spindle, 20 rpm, room temperature), were added to the aqueous starch dispersion over a period of about two hours, to produce an oil-in-water emulsion.

The amount of oil in the resultant emulsion was about 4% by weight. The viscosity of the milky-white size press emulsion was about 90 centipoises Brookfield (No. 2 spindle, 60 rpm, 106°F). The emulsion was stirred for about one-half hour prior to screening and pumping to a size press located between two drier sections on a commercial papermaking machine.

The paper being made on the papermachine was comprised of about 80% neutral sulfite semichemical pulp and about 20% sulfate pulp. The paper had a basis weight of about 54 pounds per ream. The emulsion was applied to the paper by a conventional vertically-oriented size press with the web of paper entering the nip of the press in straight-through fashion.

The emulsion was placed on the top of the traveling web by means of shower pipes located immediately before the nip formed by the press rolls. The underside of the web was likewise sized with the oil-in-water emulsion by applying the emulsion to the lower press roll which delivered the emulsion to the paper web as it passed into the nip formed by the upper and lower press rolls.

After impregnating with emulsion the traveling web was dried to a moisture content of 6%. An analysis of the paper showed that it contained about 1.1 pounds of oil per ream of treated paper. The oil, which represented only about 1.9% by weight of the paper, was uniformly distributed throughout the paper. The opacity of the paper with a starch sizing only was about B&L 75, and the opacity of the paper treated with the oil-in-water emulsion was about B&L 72.

The paper was later slit into appropriate widths to produce code paper for perforator tape, and the tapes performed exceptionally well when enclosed and used as coded tapes. There was no oil odor attendant with the code paper and it did not suffer from oil bleeding in use.

PAPER FOR FOOD PRODUCTS

Edible, Water-Soluble Pullulan Paper

The process of *T. Nomura; U.S. Patent 3,936,347; February 3, 1976; assigned to Sumitomo Chemical Company Limited and Hayashibara Biochemical Laboratories Incorporated, Japan* pertains to a specific paper composed mainly of pullulan fibers.

The pullulan is a linear compound that is different in molecular structure from other polysaccharides such as starch. It is obtained as a slime secreted in a culture liquor from a strain belonging to the species *Pullularia pullulans* which is an incomplete microorganism [H. Bender, J. Lehmann et al, *Biochem. Biophys. Acta,* 36, 309 (1959); Seinosuke Ueda, *Kogyo Kagaku Zasshi (Journal of Industrial Chemistry),* 67, 757 (1964)].

The pullulan is produced by a process comprising incubation of pullulan-producing strains on an aqueous medium containing as a carbon source oligosaccharides and/or dextrin, i.e., partially hydrolyzed starch having dextrose equivalent (hereinafter referred to as DE) of 20 to 70 which is obtained by acidic or enzymatic hydrolysis of starch as disclosed in Japanese Patents 44492/73 and 42894/74.

The employable pullulan-producing strains include various strains which belong to genus Pullularia such as *Pullularia fermentans* var fermentans IFO 6401, *Pullularia fermentans* var fusca IFO 6402, *Pullularia pullulans* AHU 9553, *Pullularia pullulans* IFO 6353, *Pullularia pullulans* IFO 4464, etc. The strains should be selected to obtain pullulan with suitable polymerization degree according to its specific applications.

Since the starting pullulan is high in water-solubility, the paper is quickly dissolved when dipped in water, or is destroyed when discarded in the fields, and hence, is usable as a paper for secret documents. In addition, since the starting pullulan is edible, nondigestible, and water-soluble, the paper of the process, when used as internal packaging materials for medicines in the form of powders or tablets, or for instant food seasongs, powdered tea, or coffee is not required to be removed at the time when the medicines or foods are taken or served.

Example 1: Preparation of Pullulan Fibers – *Pullularia pullulans* IFO 4464 was used as pullulan-producing strain. For the cultivation of the strains, a culture medium comprising the following composition was prepared. As a carbon source, enzymatic hydrolyzed starch syrup (DE 40) was used. Composition of the culture medium in percent by weight was as follows:

Carbon source	10
K_2HPO_4	0.2
Peptone	0.2
NaCl	0.2
$MgSO_4 \cdot 7H_2O$	0.04
$FeSO_4 \cdot 7H_2O$	0.001

100 ml of the medium with the above composition was placed in a 500 ml flask and sterilized by heating according to the usual method. Strains obtained by slant-cultures for 2 days and then by seed-culture for 2 days were inoculated on the medium placed in the flask, and then was cultivated in a rotary shaker at 27°C, with pH 7.5 under a vigorous stirring for 5 days.

The resulting viscous liquor was centrifuged to precipitate and remove the cells, decolored by addition of 0.5% by weight, based on the weight of the liquor, of active carbon, and then filtered to obtain a transparent colorless liquid. To this liquid, 2 times the volume of the liquid of isopropanol was added with vigorous stirring, whereby a white fibrous pullulan precipitate was deposited.

This precipitate was recovered by filtration to obtain 65%, based on the starting starch, of pullulan (after drying) having an average molecular weight of 250,000. The thus obtained wet pullulan fibers are referred to as pullulan fibers A, hereinafter.

The pullulan fibers A were decreased in water content to 15% and then powdered and the resulting powder was extruded, by use of a heat-kneading type spinning machine, through fine orifices at a temperature of 100° to 160°C under a pressure of 100 to 150 kg/cm^2 and stretched to obtain fine pullulan fibers, which were then cut to a length of 5 to 10 cm to prepare short pullulan fibers. The thus prepared short pullulan fibers are referred to as pullulan fibers B, hereinafter.

Example 2: Pullulan Paper — (a) The pullulan fibers A obtained in Example 1 were dispersed in a 90% aqueous acetone solution to a concentration of 0.5%, and the resulting dispersion was sufficiently stirred to prepare a homogeneous dispersion. This dispersion was subjected to a Fourdrinier machine.

That is, the dispersion was flowed on a wire screen and sufficiently drained, with vibration, to a concentration of about 9% to form a paper web. The paper web was repeatedly dehydrated by means of squeeze rolls to a concentration of about 39%, thereafter dried with drying rolls and hot air at 30° to 40°C and finally dried at a temperature of 80°C to obtain a paper having a water content of 15%. The thus obtained paper had a thickness of 0.1 mm, was high in flexibility, had a surface similar to that of a Japanese paper, was suitable for writing with pens and pencils, and was not penetrated nor blotted with inks.

(b) A mixture comprising 2 parts of the pullulan fibers A obtained in Example 1 and 1 part of wood cellulose pulp was suspended in a 80% aqueous acetone solution to a concentration of 0.3%, and the resulting suspension was sufficiently stirred to prepare a homogeneous suspension.

This suspension was subjected to papermaking treatment according to the procedure of Example 2(a) to obtain a paper, which was then treated with final calender rolls at 80°C to make the surface thereof somewhat glossy.

The thus obtained paper had a thickness of 0.1 mm and showed such excellent properties as an elongation of 1.5 in lengthwise direction and 2.6 in widthwise direction, a tearing strength of 2.5 kg/cm^2, and a folding strength of 980 in lengthwise direction and 560 in widthwise direction, was high in flexibility, had a smooth but less glossy surface, and was similar in texture to a Japanese paper. The paper was favorable in printing ink receptivity, and was not blotted nor penetrated with inks. When charged in stationary cold water, the paper lost its original shape within 8 seconds and formed a suspension of wood pulp.

(c) The pullulan fibers B obtained in Example 1 were sprayed with one-third the weight thereof of a 80% aqueous maltitol solution, and then allowed to absorb the solution for 10 hours. Thereafter, the fibers were homogeneously suspended in 250 times by volume, based on the weight of the fibers, of pure methanol to prepare a homogeneous suspension.

This suspension was sufficiently drained and dehydrated in the same manner as in Example 2(a), and the resulting paper web was sufficiently dried with air, gradually heated, and then subjected to final rolls at 80°C to obtain a paper having a water content of 14%.

The thus obtained paper was relatively flexible, had a thickness of 0.8 mm and showed such excellent properties as a folding strength of 1,150 in lengthwise direction and 630 inch widthwise direction, and a tearing strength of 2.1 kg/cm^2. Further, the paper was quite quickly and highly soluble in water such that a small bag containing a cocoa powder for a cup of cocoa was prepared by use of the paper, and charged into 130 cc of hot water, and then the water was stirred, whereby the small bag immediately dissolved and dispersed in the water to give a cocoa liquid containing no residue of the small bag.

(d) A mixture comprising 1 part by weight of the pullulan fibers B obtained in Example 1 and 1 part by weight of wood pulp was homogeneously suspended in 150 parts by volume of 90% aqueous isopropyl alcohol solution. The homogeneous suspension thus obtained was subjected to the same papermaking treatment as in Example 2(a), except that the paper web was dried with drying rolls kept at 55° to 90°C and passed between calender rolls under a slight pressure.

The resulting paper was relatively bulky but had a smooth surface with less nap and was elastic. When dipped in water, the paper was easily swelled and destroyed and when subjected to printing or writing, the paper showed excellent ink receptivity and was smoothly printable or writable without penetration of inks.

Reinforced Waterproof Paper

T. Uchida; U.S. Patent 3,912,835; October 14, 1975; assigned to Nichimen Company, Limited, Japan describes a process of reinforcing paper such as corrugated cardboard, pasteboard and other thick or thin paper of various kinds.

This comprises the steps of drying paper to a moisture content of less than 5%, dipping the paper for about 1 second to about 20 seconds in a solution made by dissolving between about 1 and 20% by weight of a resin selected from the group consisting of polystyrene, acrylonitrile-styrene copolymer, acrylic resin and acrylonitrile-butadiene-styrene resin in a hydrocarbon halogenide solvent, and thereafter drying the paper to obtain a reinforced paper.

In the reinforcing process the solution penetrates smoothly into the paper to be processed without need of penetrating agent so that the paper is sufficiently impregnated with the solution within a very short time. Since the reinforced paper is impregnated and coated with the plastics, the paper has excellent waterproof property and strength. Thus, the reinforced paper is particularly useful when it is utilized as a container for frozen foods, fresh fish etc.

Example: A liner and a corrugated sheet made as materials of a corrugated cardboard are dried by a drying machine to water content of less than 5%. Then the dried liner and corrugated sheet are dipped for five seconds in a solution made by dissolving 10% by weight of an acrylic resin in a hydrocarbon halogenide solvent. The paper thereafter is dried through a drier to obtain a desired reinforced paper. Properties of the reinforced paper produced in the above described manner are shown in the tables on the following page.

It will be apparent from these tables that the paper treated by the process is superior in burst factor, breaking strength, ring crush and sizing degree to the untreated paper. It will also be noted that the paper reinforced by the process does not entirely lose its properties even after soaking it in water for three hours.

Strength Test for Liners

	Untreated Material	Treated Material: Before Soaking in Water	Treated Material: After Soaking in Water for Three Hours
Burst Factor (kg/cm^2)	4.46	7.54	1.29
Breaking strength	2.41	3.40	0.602
Ring crush (kg)	45.1	66.2	-
Sizing degree (sec)	130	858	-
Water absorption (%)	-	-	33.1

Strength Test for Corrugated Sheets

	Untreated Material	Treated Material: Before Soaking in Water	Treated Material: After Soaking in Water for Three Hours
Burst factor (kg/cm^2)	1.93	4.29	0.85
Breaking strength	1.41	2.65	0.540
Ring crush (kg)	23.4	55.9	-
Sizing degree (sec)	4	55.0	-
Water absorption (%)	-	-	32.4

Fibrous Food Casing—Paper Web Bonded with Resin Mixture

The paper webs utilized in fabricating fibrous food casings are commercially formed on conventional papermaking machines where long fibers are beat into a slurry and then formed into webs and dried. The fibrous or paper web is then passed through or contacted with a dilute viscose solution and again dried whereupon the cellulose is partially regenerated.

The cellulose in the viscose can be further regenerated by passing the paper web through dilute acid solutions and finally a washing solution. The thusly treated paper web is then dried and cut into unit widths from a master roll to provide lesser rolls which are then utilized in the manufacture of fibrous food casings.

It has been found that the fibrous food casings obtained exhibit wide variations in extensibility although they are, ostensibly, the same size. In an effort to minimize or eliminate the wide variations of extensibility in the transverse direction of the master roll paper web and thereby improve the uniformity of extensibility of fibrous food casings produced therefrom in order to ultimately obtain food products which would exhibit dimensional uniformity alternative materials for bonding the paper webs were investigated other than the commercially employed acid-regenerated viscose.

H.S. Chiu and F.T. Kambara; U.S. Patent 3,484,256; December 16, 1969; assigned to Union Carbide Corporation found that use of a resin mixture comprising a cationic thermosetting resin and a polyacrylamide resin as the bonding agent for the paper web results in a synergistic effect in improving the overall properties of the paper web.

The cationic thermosetting resins and the polyacrylamide resins comprising the bonding agent mixture are each commercially obtainable and are those which can be cured to an insoluble state but which are compatible with water; that is, they are water-soluble or water-dispersible in the uncured state. Examples of the cationic thermosetting resins that can be used are the reaction products of an epichlorohydrin and polyamide, a modified melamine and formaldehyde and a modified urea and formaldehyde.

The polyacrylamide resin can be either the ionic or nonionic form. Although both forms of the polyacrylamide resin are generally compatible with cationic thermosetting resins, the nonionic form of polyacrylamide resin is preferred since it has been found to be compatible with a cationic thermosetting resin over a wider range of concentrations than the ionic form of polyacrylamide resin.

The bonding agent mixture comprising a cationic thermosetting resin and a polyacrylamide resin can be utilized to treat paper webs either by adding the resinous mixture to the fibrous slurry prior to formation of the paper webs, or by treating the dried or semidried, formed paper web with the resinous mixture in solution, such as by saturating the formed paper webs, or by a combination of these methods. Preferably, the formed, semidried paper web is saturated with a mixture of the resins in solution.

For optimum results it has been found that up to 4.5% by weight of the cationic thermosetting resin can be incorporated in the paper web at any time prior to subsequent impregnation of the paper web with viscose.

Generally, the amount of polyacrylamide resin which can be employed can be from about 0.1 to 1.1% by weight based upon the dry weight of the paper web. The fibrous food casings produced in accordance with these methods have been found to exhibit improved extensibility characteristics and retention of uniform dimensions—factors which are of importance to commercial food and/or meat packers.

Regardless of the methods employed to produce the fibrous food casings, the thusly treated paper webs are completely embedded in the casing wall; that is, subsequent impregnation of the paper webs with viscose results in a fibrous, cellulosic food casing where the treated paper web is entirely coated by and with the viscose to obtain a regenerated cellulosic food casing having a fibrous web embedded therein.

Paper webs were obtained from a commercial papermaking apparatus and were treated by saturating them with an aqueous solution containing a resin mixture comprising a cationic thermosetting resin (polyamide-epichlorohydrin) and the nonionic form of a polyacrylamide resin. Some of the paper webs were treated solely with a cationic thermosetting resin (polyamide-epichlorohydrin) while others were bonded by acid regenerated viscose and used as a control. The tensile properties of the thusly treated paper webs were determined and the burst pressures of fibrous food casings formed therefrom were also measured. The results obtained are set forth in the table on the following page.

The values listed in the column labeled Dry Break Strength were obtained by placing the dried, treated samples in the Instron testing apparatus. The values listed in the column labeled Wet Break Strength were obtained by first soaking the treated samples in water until they were thorougly wet before placing them in the Instron testing apparatus. The values listed in the column labeled 6% NaOH Wet Break Strength were obtained in the same manner as those for Wet Break Strength except that sodium hydroxide (NaOH) was added to the water until a NaOH concentration of about 6% was obtained. At this level of NaOH concentration the solution closely simulates the conditions a paper web would encounter when it is coated with viscose in accordance with commercial casing manufacturing processes since in these processes, viscose normally has a NaOH concentration of about 6%. The values obtained from these tests are expressed in grams per inch width of sample paper tested.

Resin content		Tensile properties of paper webs (gm./in.)						
Thermosetting resin (polyamide-epichlorohydrin)	Polyacrylamide resin, non-ionic	Dry break strength		Wet break strength		6% NaOH wet break strength		Casing burst pressures (mm. Hg)
		MD	TD	MD	TD	MD	TD	
3.13	0	2,644	2,138	888	810	393	383	512
3.77	0.45	4,593	3,645	1,041	943	407	363	518
3.86	0.77	4,039	3,645	1,160	1,099	521	481	499
3.89	1.08	4,264	3,724	1,133	1,031	507	460	501
Control (bonded with acid regenerated viscose)		5,524	4,376	1,315	1,146	329	270	447

The designations, MD and TD denote the machine direction and the transverse direction, respectively, of the formed, fibrous paper webs. The values listed under the column labeled Casing Burst Pressures are expressed in millimeters of mercury and were obtained by closing one end of the formed tubular casings and inflating them with air through their open ends until the casings ruptured.

A comparison of the results reveals that the tensile properties of the paper webs treated with the resin mixture are significantly higher than those obtained from the paper webs treated with only the cationic thermosetting resin. Similarly, the burst pressure of fibrous food casings utilizing paper webs bonded with either the polyamide-epichlorohydrin resin alone or with the resin mixture are significantly higher than the fibrous food casing formed from the paper web as the control.

FILTERS

Resin-Chrome Oxide Impregnated Oil Filter Paper

Present day oil filter papers, must be full-flow papers, that is, they must pass all of the oil used in an engine. Such filters have for a long time been made of a resin-impregnated paper. From the point of view of economy, the beater addition method is the best method of making the resin impregnated filter paper.

The principal drawback of the beater addition method is that considerable resin may be wasted because it will not have been completely precipitated onto the fibers and will be carried away by the water at the papermaking machine wire or cylinder.

C.A. Charron; U.S. Patent 3,539,446; November 10, 1970; assigned to Rochester Paper Company found that this resin loss can be substantially reduced if the resin-fiber containing slurry is treated with the addition of either chrysotile asbestos or chrome oxide prior to the formation of the paper web. The pulp should be one of a type normally used for making filter papers, for example, Hercules #46 cotton linters, or it may be in certain instances an unbleached sulfate from southern pine wood, or various bleached or unbleached northern or southern soft woods.

The advantage gained by using cotton linters is that a highly porous paper will be obtained even though the pulp is beaten to a lower freeness than most wood pulps. The precipitant is then advantageously added to the pulp in the washer or beater and in sufficient quantity to render the pulp acid to the extent that it has a pH of from 2.5 to 3.5.

After addition of the ingredients, the pulp containing slurry is then fed to a refiner such as a Jordan. It is discharged from the refiner into a mix-box where a suitable amount of resin, one of the beater addition types, is added in an amount such that it will comprise about 15 to 45% by weight of the final product with a retention of at least 55% due to the inclusion of the chrysotile asbestos or chrome oxide.

The amount of chrysotile asbestos ranges from a small but perceptible amount to 5% by weight of the total solids content of the slurry. The amount of the chrome oxide ranges from a small but perceptible amount to 2% by weight of the total solids content of the slurry. The pulp-resin slurry with the small amount of chrysotile asbestos or chrome oxide added is then fed to a pump by which it is lifted to the head box of a papermaking machine. The paper is then produced as in a conventional process.

An applicable phenolic resin is Synco Grade 832 (Snyder Chemical Company), phenol-formaldehyde resin emulsion. The mol ratios of phenol to formaldehyde should preferably be in the range of 1:1 to 1:2.5.

Example 1:

	Percentage by Weight
Cotton linters	39
High alpha kraft (bleached)	18
Chrysotile asbestos fibers	5
Water dilutable phenol-formaldehyde resin	38
	100

The fibrous materials were dispersed in a beater and dumped into a stock chest. The resin was slowly added as the mixture was only agitated mildly so as to preclude foaming before exhaustion. The contact time was at least 15 minutes. The water temperature was 54°F and the pH was lowered to less than 3.5. Another contact time of 15 minutes was required to insure maximum precipitation. The resultant slurry was then utilized to produce a paper by employing a conventional paper machine.

A slurry containing the following ingredients was utilized to produce a paper in accordance with the procedure described in Example 1. The amount of rayon contained in the cotton linters was based on the weight of the linters.

	Percentage by Weight
Cotton linters containing 10% by weight of 3 denier rayon ⅜" cut	61
Chrome oxide	1
Water dilutable phenol-formaldehyde resin	38
	100

The characteristics of the paper made in the above examples are as follows.

	Example 1	Example 2
Basic weight in pounds per ream (24" x 36", 500 sheets)	140	60
Thickness in inches	0.048	0.028
Stiffness (Gurley) MD, mg	7,120	-
Stiffness (Gurley) CD, mg	3,110	-
Tensile, lb/in MD	20.3	9.3
Tensile, lb/in CD	8.9	3.8
Mullen, lb/in^2	17	15
Air permeability, cfm/ft^2 at 0.5" pressure	27.5	75
Depth of corrugations, inches	0.015	0.012
Resin content, percent	28.0	26.0
Resin retention, percent	69.6	57.0

To illustrate the advantages obtainable, a comparison was made between the product of Example 1 and a conventionally produced paper having a phenolic resin retention of 45%. The comparison concerned dust (in air) filtration characteristics. The papers otherwise had nearly identical physical properties.

Sample	Capacity, grams	Efficiency, percent
Example 1	77.0	99.60
Conventional paper	61.0	96.44

Capacity is amount of dust in grams retained on a 900 square inch surface sample up to a point where pressure drop across the material thickness was 10 inches of water. The efficiency refers to percentage dust retained in filter as a function of the total amount fed (feed concentration was 0.10 gram dust per cubic foot air; feed velocity 35 feet per minute).

Filter Paper for Oil Filters

T. Ochiai; U.S. Patent 3,640,839; February 8, 1972; assigned to Tokyo Roki Kabushiki Kaisha, Japan provides a filter paper for oil filters comprising a coarse web capable of passing particles of granular size of about 40 microns maximum diameter and a plurality of dense web strips superposed in parallel upon the coarse web and combined therewith at fixed intervals, the dense web being capable of passing particles of granular size of about 15 microns maximum diameter.

The fiber mixing proportion of the coarse web is by weight from 40 to 60% of cotton fiber and from 60 to 40% rayon (from 1.5 to 7 denier thick and from 5 to 10 mm staple length), while the dense web consists of 100% cotton fibers.

When oil which is being filtrated is circulating through this filter paper, the double layer parts will carry out the cleansing function so that minute particles in oil will be removed, while the single layer parts will carry out the screening function so that particles which may have bad influence on the wear of the machine will be removed.

When the double layer parts are closed by dust and dirt, the single layer parts will also become closed by them, so that not only the filtering function will not be lowered as a whole, but also the cleansing efficiency for oil will be raised compared with that at the beginning of use. Furthermore, with this construction, the service life of the filter paper will also be extended. The plane of the dense web may be of any shape.

Figures 9.1a through 9.1d show filter papers **10, 10a, 10b** and **10c**, which are made by superposing relatively dense wet webs **30** upon relatively coarse wet webs **20** and combining them by pasting or the like. The filter paper is so constructed that it comprises, as a vertical sectional view, double layer parts in which a dense web is superposed upon a coarse web. Some filters may also have single layer parts of a coarse web.

In 9.1a the layer of dense web is arranged in parallel longitudinal strips. In 9.1b very thin dense web parts **32** extend between thick dense web strips **31** and upon the coarse web **20**. In 9.1c and 9.1d the dense web **30** is regularly punched forming holes **33**; 9.1d has an additional very thin dense web layer **32**.

In the constructions of (b) and (d), relatively thick dense web parts are connected with the very thin dense web parts **32** and all dense web parts are combined by being pasted upon the coarse web **20**. Therefore, the filter paper of these constructions will hardly be torn off even if the filter is bent by the pleat machine. Furthermore, the strength of the filter is increased by these structures. In order to keep high degree of filtration for oil, the thickness of the web parts **32** should be as small as possible.

The ratio of planer exposed surface area of a plurality of dense webs to that of exposed coarse web is in the range of 10:5 to 10:10. If stress is laid upon the filtering efficiency of the filter paper, the desirable ratio is 10:6. If greater value is placed on the service life, the ratio should be 10:10. When a filter element is constructed by applying this filter paper, it is important that the filter paper should be so arranged that oil may flow through from the side of the coarse web **20**.

In order to make not only the coarse web but also the dense web for the filter paper by the use of the Fourdrinier machine **50** (Figure 9.1e), flexible covers are applied on the surface of the moving endless belt of wire cloth (the Fourdrinier wire) or by which wet dense web is made, so that the mesh of the wire may be closed partially by the covers.

As the flexible cover, for example, an adhesive sheet, a narrow steel plate or the like can be used. These covers are fitted on the moving belt in rows at fixed intervals. Also the surface of the wire may be covered with a cloth upon which holes are regularly formed.

A fiber-water mixture, which is the starting material, is used in which the fiber mixing proportion is 50% of cotton fiber and 50% of rayon fiber.

FIGURE 9.1: FILTER PAPER FOR OIL FILTERS

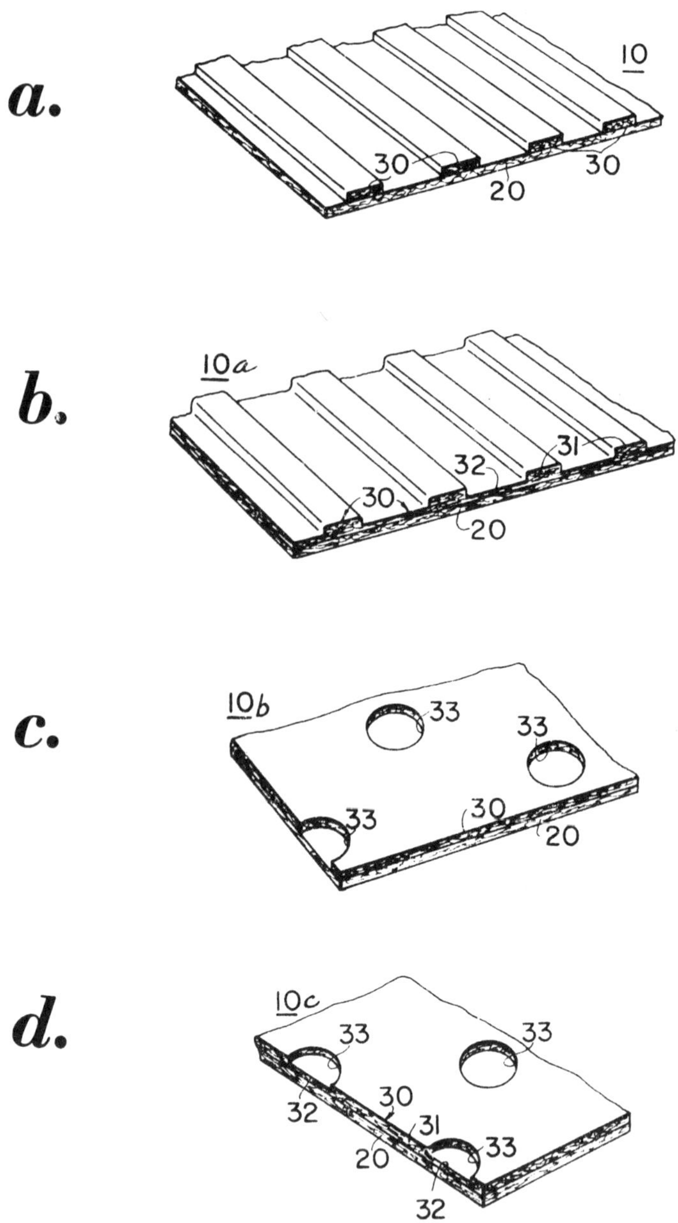

(a)-(d) Partial Perspective Views of Filter Papers on Which Dense Web Is Superposed on Coarse Web.

(continued)

FIGURE 9.1: (continued)

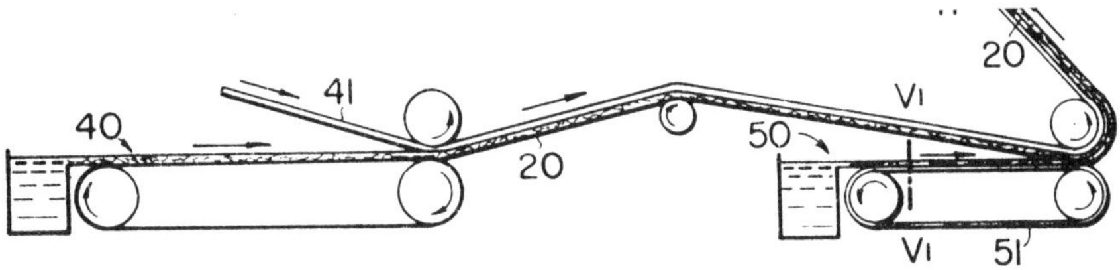

(e) Fourdrinier Machine

Source: U.S. Patent 3,640,839

This mixture is applied to the papermaking machine **40** and the wet web **20**, through which the granules of a maximum 40 microns diameter may pass, is made in accordance with the conventional method.

The fiber-water mixture where the fiber is 100% cotton is applied to another papermaking machine **50** having a plurality of thin steel plates which are 6 mm wide and spaced to the extent of 10 mm. These plates are fitted in the lengthwise direction around the wire mesh **51** so that a wet web **30**, through which granules of a maximum diameter of 15 microns may pass, is made. The wet web **30** made by arranging the liquid surface of the fiber-water mixture is combined with the other web **20** attached to and carried by woolen fabric or felt **41** so that the filter paper is made.

Microporous Fibrous Filter Material

D.B. Pall and C.A. Keedwell; U.S. Patent 3,573,158; March 30, 1971; assigned to Pall Corporation describes a method for manufacturing microporous materials having three or more strata per layer and one or more layers, each layer separately prepared and bonded or laminated together, of which at least one layer is microporous, and has a pore diameter of less than 25 microns.

The process comprises flowing upon a foraminous support a slowly flocculating dispersion in a dispersing liquid of fibrous material. It is composed of both long and short fibers which may also be of different diameters and preferably have an average length:diameter ratio of from 50:1 to 5,000:1 and a diameter within the range from 0.01 to 10 microns, and optionally nonfibrous particulate material, and a binding agent.

A supernatant layer of the suspension on the support is formed with at least a proportion of separate fibers. A thin base stratum of predominantly separate fibrous material having long and short fibers lying almost entirely in planes approximately parallel to the plane of the layer is deposited on the foraminous support.

. nen a thin microporous stratum composed of predominantly short deflocculated fibers, also lying almost entirely in planes approximately parallel to the plane of the layer, and having a maximum pore diameter of less than 25 microns, is deposited by flowing through the support not over 20% of the dispersing liquid, at a pressure differential of less than 12 inches of water, while breaking up any large clumps of fibers that may form.

The remaining supernatant liquid of the dispersion is then flowed through the two thin strata by gravity draining, or by applying a direct pressure to it, or by applying a vacuum to the underside of the foraminous support. The remaining flocculated fibrous material of the dispersion is deposited on the thin strata, with a proportion of the fibers in the added third stratum oriented so as to extend in a direction across the plane of the layer at an angle thereto greater than about 30°, sufficient to space the fibers in the third stratum at an average pore diameter below about 25 microns and a voids volume of at least 75%. The microporous material can then be stripped from the support, or the support can be allowed to remain, as a substrate, for greater strength, or for fabricating special types of structures.

The process also attacks the problem of uniformity in laydown, referred to heretofore in the discussion of the manufacture of paper, by ensuring uniform distribution of flocculated and deflocculated fibrous material in the suspension thereof prior to laydown on the screen.

This is done by preventing deposition of large flocs, before they can be laid down. The slurry is deflocculated at the time of the initial gravity laydown of the separate fibrous material on the substrate, or wire screen, and thereafter the distribution of fibrous material in the slurry, which is slowly flocculating, is maintained substantially uniform prior to final laydown. In this way, the final microporous layer produced has a substantially uniform thickness and density, and is substantially free of light spots.

The process is described with reference to the apparatus of Figure 9.2, which comprises an endless Fourdrinier wire mesh belt **10** which is driven by two spaced apart rolls **12** and **14** over a plurality of guide rolls **15**. The belt travels between closely abutting side frames **11**, which engage the edges of the belt in a fluid-tight seal, and with the belt as the moving bottom constitute a reservoir to retain a slurry of particulate material on the belt to the desired depth.

A head box **16** for delivering a slowly flocculating slurry of particulate material to the wire belt is positioned in close proximity to the roll **14** which is the breast roll and belt **10**. The head box includes a paddle wheel **2** which creates a shearing action at the head of feed passage **3** through which slurry is led to the belt **10**.

The passage has a long horizontal end portion **4** or apron which transfers the slurry to the belt **10**, and transit time through this is long enough to permit flocculation to begin. By the time the slurry reaches the belt, it is partially flocculated.

A series of gravity drainage boxes **18** and vacuum or suction boxes **20** and a series of dryers **27** are disposed in close proximity to the wire belt **10**. Immediately prior to the vacuum boxes **20** and between the gravity boxes **18** and

FIGURE 9.2: MICROPOROUS FIBROUS FILTER MATERIAL

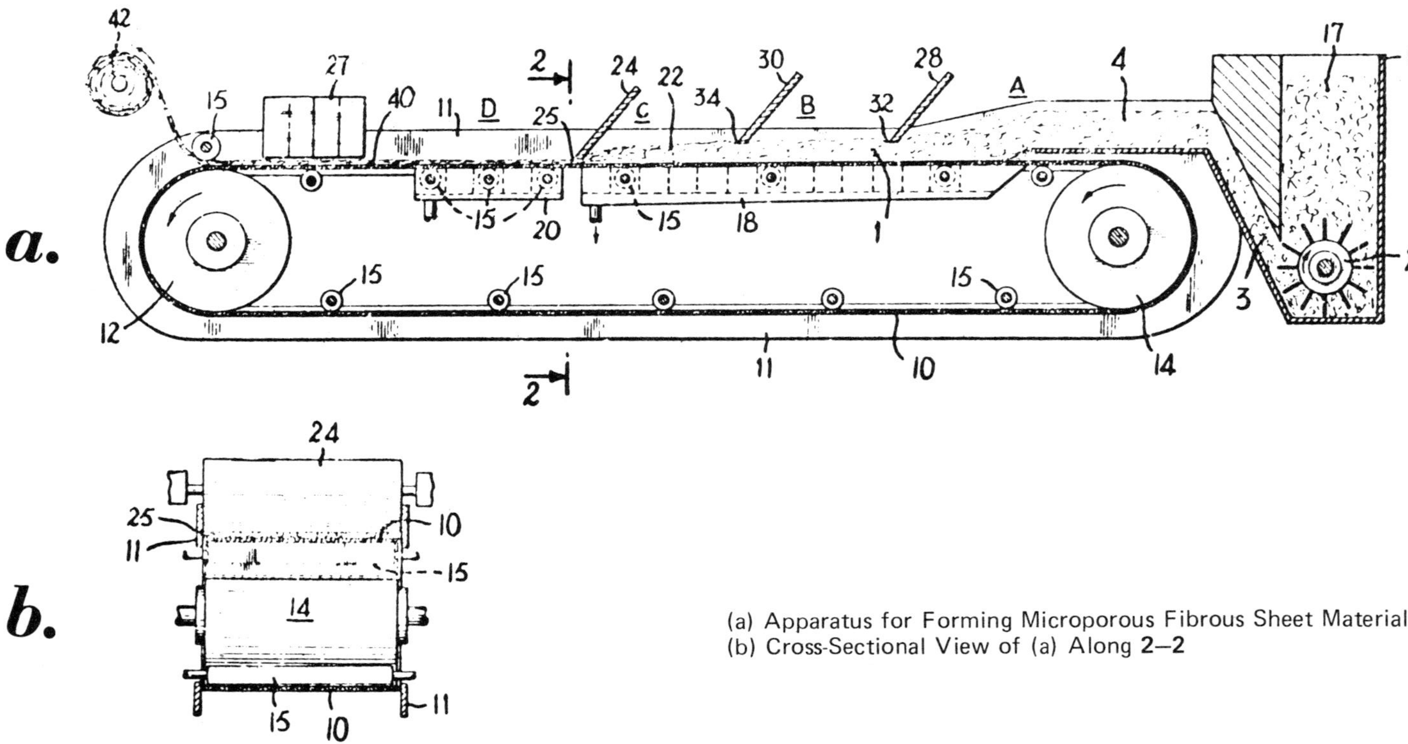

(a) Apparatus for Forming Microporous Fibrous Sheet Material
(b) Cross-Sectional View of (a) Along 2–2

Source: U.S. Patent 3,573,158

the vacuum boxes **20**, a primary deflocculating blade **24** in the form of a stainless steel blade having a thickness of 0.010 to 0.015 inch is adjustably disposed above and across the wire belt **10** to define a gap **25**, from about 0.01 to about 0.5 inch and inclined in the direction of flow at an angle to the wire belt within the range from about 30° to about 45°.

The gap is adjusted so as to restrict fluid flow, and build up a predetermined head or depth of slurry **1** on the belt in the relatively quiescent zones **A, B, C** of gravity drainage, so as to control gravity settling at a desired rate there, and also reduce the depth of the slurry on the wire belt over the vacuum box to the desired depth, and break up large flocs and clumps of particles in the slurry before the slurry is drawn down on the support over the vacuum boxes **20**.

Disposed between the primary deflocculating blade **24** and the head box **16** in the quiescent zones **A, B, C** are auxiliary deflocculating blades **28** and **30**. The auxiliary deflocculating blades **28** and **30** are disposed above and across the wire belt **10** to define gaps **32** and **34**, respectively.

The gaps defined by blades **28** and **30** are from about 2.5 to about 0.2 inch high, and the blades are inclined with flow at an angle to the wire belt within the range from about 30° to about 45°. As indicated hereinbefore, the auxiliary deflocculating blades are especially important in the case of a slow drainage slurry, to break up large flocs in the slurry in this zone.

In carrying out the method using the apparatus as shown feed stock or slurry **17**, such as an aqueous slurry of asbestos fibers, is fed from the head box **16** onto the wire mesh belt **10** in zone **A** where it is retained by the blade **28**, within the reservoir created by sides **11** of a depth of, for example, about three inches, and then passed through gap **32** into zone **B**, where depth is controlled by blade **30** to, for example, about 2 inches, initially, and then through gap **34** into zone **C**, where depth is controlled by blade **24**, to, for example, about 1.5 inches, and finally through gap **25** into zone **D** over the vacuum boxes **20**.

In zone **A**, about 5 to 10% of the slurry fluid drains quickly by gravity through the wire mesh belt, to form the initial stratum. The slurry is only partially deflocculated, if at all, at this stage. Thereafter, flocs present in the slurry which are larger than gap **32** are broken up, as the slurry passes through gap **32** into zone **B**.

In zone **B**, another 5 to 10% of the slurrying fluid of the dispersion, which has now become from 80% to 100% flocculated, gravity drains through the wire mesh belt forming the first portion of the intermediate stratum **22**. It then passes beneath the blade **30** into zone **C**, and by the end of zone **C**, from about 15 to about 75% of deflocculated particulate material in the flocculated slurry, which is substantially free of large flocs, has gravity-settled on the mesh belt to form an intermediate stratum **22**.

The supernatant slurry of particulate material above the initial layer is passed through the gap **25** beneath blade **24**, and as the slurry flows through the gap and enters zone **D** the turbulence on the head box side breaks up large flocs in the slurry and reduces the depth of the slurry to the approximate width of the gap.

The fibrous material distribution in the slurry over the vacuum boxes **20** is now uniform. Immediately thereafter, while the slurried flocculated fibrous material is still uniformly distributed, and before large nonuniform flocs can form, the remaining supernatant slurrying liquid is drawn off over the vacuum boxes **20**, through the initial and intermediate strata, and through the mesh belt and the remaining flocculated fibrous material is laid down as the top stratum of the layer, thereby forming a tristratal fibrous sheet material **40** substantially uniform in thickness and density. The sheet is then passed under the dryer **27**, and thereafter stripped from the mesh belt, and rolled up at **42**.

The barrier means or deflocculating blade or blades is ordinarily a thin plate of wood, glass, metal, such as stainless steel, or plastic, and have a straight edge disposed across the wire mesh belt and parallel thereto. These can be provided with adjusting screws to vary the height and thickness.

Such blades, often referred to as doctor blades, should be at least as wide as the support to ensure that substantially all slurried material must pass under the blade. In addition, the length of the blades is dependent upon type and rate of drainage of the slurry, and the width, pore size and speed of the wire mesh support.

Example 1: The apparatus shown was used in the following way to prepare a fibrous sheet of uniform thickness and density, suitable for use as a filter medium by coating an alpha-cellulose-hemp paper substrate with asbestos fibers.

The deflocculating blades **24, 28** and **30** were stainless steel strips, 42 inches in length and about 0.010 to 0.015 inch in thickness. The primary deflocculating blade **24** was positioned between the vacuum box and head box about nine-tenths of the distance from the head box, and disposed above and across the wire mesh belt **10** at an angle of 40° to the mesh belt, and defined a gap about 0.024 inch high over the mesh belt.

The auxiliary deflocculating blades **28** and **30** were positioned between the blade **24** and the head box about 8 inches from each other, blade **30** being about 8 inches from blade **24** and blade **28** being positioned at the beginning of belt **10**. Blades **28** and **30** were inclined at an angle of 40° to the belt and formed a gap of about 0.375 inch above the belt. The wire mesh belt contained openings 0.008 inch x 0.012 inch.

A slurry of crocidolite asbestos was prepared by mixing 1,500 gallons of water with 75 pounds of crocidolite asbestos fibers averaging 0.01 micron in diameter and 1 to 30 microns in length. The slurry of asbestos was poured into the head box, and was fed therefrom at a rate of about 9 gallons per minute onto the substrate, an alpha-cellulose and hemp paper, with 25% polyvinylidene chloride as a binder, weight 4 grams per square foot, thickness 0.005 inch, supported on a wire mesh belt traveling at a rate of 9 feet per minute.

In zone **A**, the slurry depth was 1.5 inches, a portion of the water drained quickly through the belt, and there was formed an initial stratum 0.001 inch thick on the substrate, from about 5% of the fibers present.

The slurry of asbestos on the substrate then moved beneath blade **28** through the 0.375 inch gap between the blade and the mesh belt, which broke up the few flocs that had formed, into zone **B**. In zone **B**, the slurry depth was 0.75 inch. The slurry passed beneath blade **30**, where the large flocs were broken up, and into zone **C**, where the slurry depth was 0.375 inch. A portion of the water continued to gravity-drain through the substrate in zones **B** and **C**, and about 15% of the asbestos fibers in the slurry had gravity-settled on the substrate by the end of zone **C**, within about 25 seconds, to form an intermediate stratum about 0.002 inch thick.

The substrate, coated with the initial and intermediate strata and the remaining water and asbestos fibers of the slurry, was passed beneath the blade **24**, in the course of which the large flocs present broke up, and the combined depth of the water and asbestos slurried therein and the thickness of the initial and intermediate strata of asbestos plus the subsrate was reduced to about 0.025 inch.

In zone **D**, the slurry and initial layer of asbestos were then passed over the vacuum box, where, before large flocs could reform, the remaining water was drawn through the subsrate and mesh belt, depositing the remaining flocculated asbestos fibers on the initial layer of fibers in a manner such as to produce a top stratum thereon which had a substantial proportion of fibers extending across the plane of the layer at an angle of at least 30°.

The fibrous coated substrate was of substantially uniform thickness, about 0.012 inch. Thereafter, the fibrous layer and substrate was passed under the dryers, dried and then removed from the wire mesh belt.

The final product was an asbestos-coated paper sheet having a maximum pore diameter of about 0.35 micron, an average pore diameter of about 0.1 micron; a water permeability of about 0.5 gallon per minute per square foot at an applied pressure differential of about 1 pound per square inch and a voids volume of about 91%.

The dried sheet had a substantially uniform thickness of about 0.014 inch with an asbestos layer about 0.009 inch thick; and a substantially uniform density of about 18 pounds per cubic foot; and was particularly suitable as a filter medium.

Microscopic inspection of a cross-section of the material showed three microporous strata. In the first two strata, nearly all of the fibers were in planes approximating the plane of the layer, and in the top stratum a proportion of the fibers were oriented at an angle of 30° or more to the plane of the layer. In place of the crocidolite asbestos fibers, the particulate material could be glass fibers, diatomaceous earth, glass fibers plus diatomaceous earth or cellulose fibers.

Suction Molded Pocket Type Filter

P.B. Burnham; U.S. Patent 3,622,446; November 23, 1971; assigned to Keyes Fiber Company provides a suction molded filter unit for filtering contaminants from streams of gas, and a method for manufacturing the unit which involves preparing an extremely thin aqueous slurry preferably having less than 0.4% by weight of primarily synthetic fibrous material.

The fibrous material includes no more than about 30.0% by weight of wood pulp, a large amount of at least one member of the group consisting of olefin, acrylic and modacrylic fibers and fibers of a copolymer of vinyl chloride and vinyl acetate, and a balance comprising fibers of rayon, a polyester, a polyamide, fiber glass or asbestos or a mixture thereof.

A layer of the fibrous material is formed by suction against an open-face forming die contoured to define the shape of the filter unit, and the layer of fibrous material is transferred from the forming die to the dryer conveyor by means of a matingly contoured transfer die.

The forming die comprises a slotted die construction with a screen covering, whereas the transfer die comprises a slotted die construction without a screen covering.

The replaceable unit for filtering contaminants from streams of gas comprises a series of pocket forming layers superimposed one upon another in layered fashion. The replaceable unit conveniently has a width dimension and a height dimension which are about identical, whereby the front and rear ends of the unit are square. The length and other dimensions of the unit may vary in wide ranges depending upon the filtering environment in which it is to operate.

Each layer is comprised of two subassembly halves, heat sealed together to provide the series of elongated pockets which are open at the upstream end of the final unit and closed at the downstream end of the final unit. The air stream in which the filtering unit is placed enters the pockets at the open end and is forced through the filtering material along the full length of the pockets.

In the assembled unit, the spacing between the pockets provides a channel into which the filtered air may flow and freely escape from the downstream end of the unit. Since each half of each layer is identical, a detailed description of only one such half is necessary. (See Figure 9.3a.)

Each subassembly half **26** thus includes an upstream side **32** and a downstream side. Viewing one of the subassemblies **26** from the upstream side, the unit has along its forward edge the lip or flange **25**. In approximately the same plane as the lip **25**, but offset slightly therefrom, is a secondary flange **36** which surrounds the remaining three sides of the subassembly.

When two such subassemblies are positioned with their upstream sides **32** face to face, the surrounding flanges **36** make planar contact so that the two subassemblies may be secured together by heat sealing or other means to form one of the pocket layers.

Each subassembly **26** further includes a series of elongated depressions or valleys **38** which are open at the upstream end and closed at the downstream end. When two subassemblies are secured together to form a layer the opposed valleys **38** from elongated pockets.

The adjacent valleys **38** in each subassembly **26** are separated by parallel ribs **40** having elongated flat tops which are coplanar with the surrounding flange **36**. When two subassemblies are secured together, the rib tops **40** may be sealed together along with the surrounding flanges **36** to insure that the individual pockets are isolated from each other.

FIGURE 9.3: SUCTION MOLDED FILTER UNIT

a.

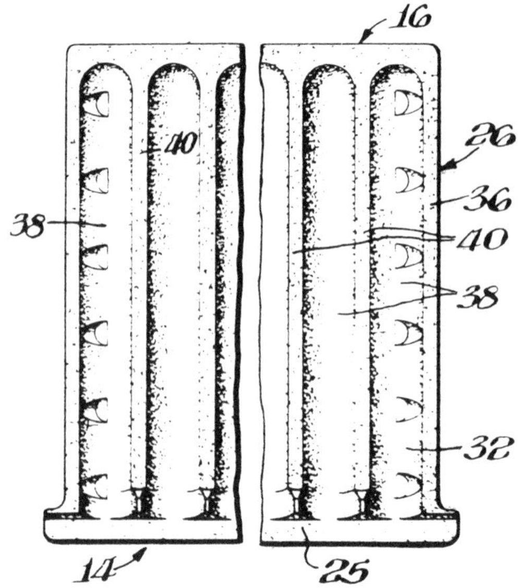

(a) Subassembly for Pocket Forming Layer

b.

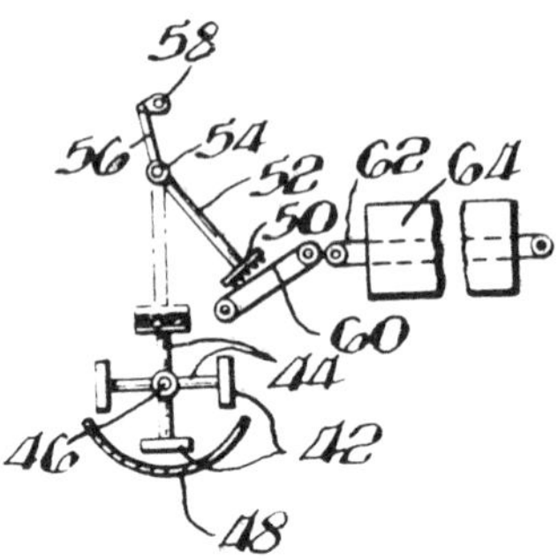

(b) Pulp Molding Machinery

Source: U.S. Patent 3,622,446

The pulp slurry is prepared by first blending approximately 2% by weight of the fibrous material with approximately 98% of water, and thereafter adding water to the blend to acquire the desired aqueous slurry containing less than 0.4% (preferably 0.05%) by weight of fibrous material.

In preparing the blend, the wood pulp is preferably first mixed with the water, the hydrophobic fibers are next blended in, and the ingredients such as rayon which blend in easily are added last. Thereafter, before the additional water is added to attain the desired density, a flame retarder may be added depending upon the ultimate use of the final filter product, as well as approximately 1% by weight cationic starch to improve the bonding of the fibrous material and 1.5% alum to reduce the pH to insure that the flame retarder acts properly on the fibrous material.

The flame retarder presently found satisfactory—and which has useful bonding properties as well—may be added to the slurry in amounts up to 50% or more on the weight of the fibrous material after the rayon and before the starch, and this advisably comprises, in approximately equal parts, a thermoplastic (not thermosetting) resinous copolymer of vinylidene chloride, and antimony trioxide pigment in an approximately 50 to 50 water suspension, such as Vi-Tard.

The preferred combination of ingredients for the fibrous material includes up to 30% by weight of wood pulp, (bleached softwood kraft), 20 to 65% by weight of olefin, acrylic and modacrylic fibers covering a wide range, as well as fibers of a copolymer of vinyl chloride and vinyl acetate containing at least 85% by weight of copolymerized vinyl chloride, and a balance comprising fibers of rayon, a polyester such as polyethylene terephthalate, a polyamide, fiber glass or asbestos or a mixture thereof. Preferably this may include 10 to 40% by weight of fibers of rayon and 10 to 40% by weight of fibers of polyethylene terephthalate.

Most specifically, as one example suitable filters have been made which include about 30% by weight of fibers of rayon having a length of up to approximately three-eighths inch wherein a minor portion of the fibers have a diameter of approximately 15 to 20 microns and a major portion have a diameter of approximately 27 to 32 microns, and about 20% by weight of fibers of polyethylene teraphthalate where the fibers have a length of approximately three-eighths inch and a diameter of approximately 15 to 20 microns.

Successful filters have been manufactured (with reference to Figure 9.3) in the following manner. First, the pulp slurry including the fibrous pulp materials of the types and in the quantities described above is prepared. This pulp slurry is delivered to the pond **48** and contained there in the conventional manner. The above described screen-covered slotted forming die **42** is immersed in the aqueous slurry and suction or vacuum at a value of from about 5 to 20 inches Hg is applied for approximately one to ten seconds.

Due to the unusually large (20%) percentage of the slotted die face which is exposed to the suction, large amounts of fluid are drawn through the screen resulting the deposition of fibrous material upon the screen. For instance, for a subassembly half **26** of generally rectangular overall shape approximately 22 x 12 inches, approximately 25 gallons of fluid pass through the die **42** at 20 inches Hg suction in a 6-second forming time.

The forming die with the layer of fibrous material thereon is then removed from the slurry and moved into position for engagement with the transfer die **50**. The layer of fibrous material on the face of the forming die at this stage is composed of approximately one part fibrous material to four parts water.

Thereafter, the matingly contoured transfer die **50** is brought into contact with the layer of fibrous material on the forming die, and lightly pressed thereagainst at room temperature for a short time such as several seconds or less. Suction or vacuum at a value of approximately 5 to 20 inches Hg is then applied to the transfer die, and the suction applied to the forming die is shut off and air pressure applied in the known fashion to effect transfer of the layer of molded fibrous material from the forming die to the transfer die.

The transfer die thereafter moves into position above the dryer conveyor **60** and transfers the layer of molded fibrous material by replacing suction with pressure in the known manner to effect release of the layer from the transfer die to the dryer conveyor.

The filter unit subassembly **26** is then conveyed through the drying tunnel **64** in the known fashion. Preferably, for one example with about 50% by weight of wettable fibers, the drying takes place for 5 to 20 minutes at an oven temperature of approximately 200° to 300°F. During the drying process, the article attains a drying temperature of about 240° to 250°F which is in the softening range but below the melting point of the copolymer of vinyl chloride and vinyl acetate.

This insures that this ingredient effects a slight bonding which is believed to increase the strength of the final article to some extent. When the article reaches the exit end of the drying tunnel **64** it is ready to be assembled with another such unit in inverted fashion to form a pocket section for eventual assembly in the replaceable filter unit.

Filtering units manufactured in the above described manner have proved highly satisfactory in actual use. The strength of the filtering material is more than adequate even for the large area of material involved with pocket or bag filters of this nature. The airflow capabilities are easily balanced against the dust-trapping capabilities of the filtering material, and these two requirements are determined by the fiber types, diameters and lengths.

For some applications, it is desirable to have higher airflow capabilities and lower dust trapping capabilities, and the ingredients of the fibrous material may be combined according to the above teachings to achieve the following requirements having desirable airflow and dust trapping capabilities throughout a wide range.

Moreover, with filters made according to this process, the air flow capabilities are maintained at the desired high level even though the replaceable unit experiences considerable clogging after long use. This is believed to be attributable to the fact that the downstream side of the article is the screen side of the article; that is, the side of the article which is in contact with the forming die **42** during the molding operation. The upstream side **32** of the article is thus fuzzier and less compact due to the manner in which the water is drawn through the article during the molding operation. When placed in use as a filter, the upstream side of the article is thus fuzzier than the downstream side, and this materially improves the anticlogging characteristics of the filter according to this process.

GYPSUM BOARD COVER PAPERS

Multi-Ply Cylinder Machine Paper

According to *R.S. von Hazmburg; U.S. Patent 3,562,097; Feb. 9, 1971; assigned to U.S. Gypsum Co.* improved gypsum boards are prepared by encasing a cast gypsum (calcium sulfate dihydrate) core within a multi-ply paper cover sheet in which at least one of the plies thereof comprises a major amount of cellulosic fibers and a minor amount of mineral fibers. The term mineral fibers as used herein means rock wool, slag wool and spun or drawn glass fibers and the term cellulosic fibers includes lignocellulosic fibers. The mineral fibers are employed in the paper cover sheet in an amount from about 2 to about 20% (most preferably 7 to 15%) by weight of the entire sheet with no single ply thereof containing more than about 43% by weight of mineral fibers.

Preferably the multi-ply paper cover sheets are made by separately dispersing a mass of cellulosic fibers and mineral fibers in water, combining the two dispersions of fibers in desired proportion and finally forming the combined fibers into a web of paper on a papermaking cylinder machine. Standard paper mill accessory equipment is suitable.

The length of the mineral fibers is important in achieving the desired results and a substantial proportion, that is about 80% or more of the mineral fibers present in the paper cover sheet, should not greatly exceed the length of the cellulosic fibers. The uncreped, multi-ply cylinder machine paper sheet has a ratio of machine-to-cross direction dry tensile strength between about 3.35:1 and 4.1:1.

Example: A seven ply paper sheet was prepared on a cylinder machine. The furnish for the exposed or liner plies was 73% newspaper and 27% magazines. The furnish for the five filler plies was basically 85% corrugated paperboard and 15% newspaper to which was added the amount of mineral wool indicated in the table on the following page.

Mineral wool was produced on an apparatus similar to that shown in U.S. Patents 2,587,710 and 2,646,593 and was treated lightly with an antidusting mineral oil and contained about 50% shot. The fibers were about 3.5 to 5 microns in diameter and a substantial proportion thereof were less than one-half inch long.

The paper was prepared in a mill having separate stock systems and the liner furnish was prepared in normal manner. The mineral wool was dispersed in a hydropulper to a consistency of about 2.4% with minimum refining to avoid undue reduction in fiber length. However, hydropulping was sufficient to insure that the mineral fibers were well dispersed and did not agglomerate into bundles or clots.

At the beginning of the papermaking operation only cellulosic fibers were used in both liner and filler plies. Mineral wool was then added to the cellulosic fibers in a filler cylinder vat in an amount of about 10 to 15% of the furnish going to that vat and soon thereafter to the balance (4) of the filler cylinder vats. As the mineral wool was blended with the filler stock at the screen head box, the stock became freer and additional water was added.

Sodium aluminate together with alum was added to the filler plies to set the size. The paper containing mineral wool dried easily. The ease with which the sheets having high mineral wool concentrations were dried is illustrated by the moisture values reported in the table below.

For purposes of analysis, a small sheet of tissue paper was periodically placed between the mineral wool-free liner plies and the filler plies so that later, when the sheet had been dried, the liner plies could be readily removed. The filler plies were then ignited at 1000°F and the ash reported as percent mineral wool in the following table. The values have been rounded off so the figures reported are consistent with the accuracy of the analysis.

Mineral Wool Paper Characteristics

Percent wool:						
Filler plies	0	10-12	15-17	18-20	28	33-46
Total sheet	0	7-9	11-12	13-14	20	27-33
Basis weight (lb/10^3 sq ft)	73.1	70.5	74.3	72.7	68.7	68.1
Caliper in 0.001 inch	20	21	21	21	21	21
Tensile strength (lb/in of width machine direction)	107	123	115	112	98.3	45
Corrected to uniform basis weight	-	127	113	113	106	49
Tensile strength (lb/in cross direction)	27	30	28	27.8	26.3	20
Corrected to uniform basis weight	-	31	27.5	28	26.3	21.5
Moisture, percent by weight	7.5	5.8	7.0	5.2	4.3	3.6
Porosity*	180	168	160	144	142	60
Shot, percent of ash	-	14.3	19.6	-	16.9	20.1

*Porosity determined using a Gurley Densometer in accordance with Technical Association of the Pulp and Paper Industry Standards T460m-49.

Referring to the table, it will be noted that the tensile strength of the paper in the machine direction and also in the cross direction when these values are corrected for the weight of the sheet increased when mineral wool in an amount between about 10 and 20% was utilized and the tensile strength did not decrease considerably until the percentage of mineral wool in the plies was greater than 28%.

A considerable amount of shot was separated from the mineral wool when it was dispersed in water in the hydropulper. The remaining shot, reported in the above table, was determined by gently brushing the residual matrix of mineral fibers from the ignited ash through a 325 mesh screen, and recovering the shot on the screen. The fibers recovered were about $\frac{1}{16}$ to $\frac{1}{32}$ inch long.

The papers of the example having plies containing 28% and 46% mineral wool were made into gypsum board using normal board forming machinery and processing. No difficulty was encountered in forming the board with either paper when used as face and back paper. Width expansion was slightly less with the mineral wool papers than with regular paper which might have been expected since the shrinkage had been less with the mineral wool paper on the paper machine. Because of the increased porosity of the mineral wool paper the board dried more easily and the temperature in the drying kiln can be 15 to 20°F lower than normal. Board made with 28% mineral wool plies was generally satisfactory but that made with 46% mineral wool was below specifications as set forth in ASTM-C-36 when measured parallel to the long edge of the board although the strength measured across the board was satisfactory.

The bond of the paper to the core was good and there was no tendency for the mineral wool paper to cockle either in the face sheet or back sheet. The lack of cockles in the face sheet is extremely desirable in a quality board and their absence on the back sheet was highly surprising as some cockles are usually present when board is made with all cellulosic fibers in the back sheet.

The term cockle is applied to a quality defect in finished board and refers to a depression having its length parallel to the machine direction which probably results from the presence of more lineal inches of paper in the depressed area than in the adjacent flat portions. The excess of paper may occur because of nonuniform formation on the paper machine or from an excess of moisture which has caused the paper to expand unevenly. Variation of moisture content may result from poor drying of the sheet on the paper machine or from some paper defect which permits moisture from the gypsum slurry to penetrate the sheet more freely in some places than in others. The increased uniformity of formation resulting from the addition of mineral fibers to the board paper enables the production of quality board with a minimum of cockles.

The porosity of paper as determined using a Gurley Densometer in accordance with Technical Association of Pulp and Paper Industry Standards T460m-49 is reported as the number of seconds required for a specified quantity of gas to pass through a specified area of the sheet; a numerically higher porosity value therefore indicates a more dense sheet. The porosity of the paper cover sheets is very important in gypsum board manufacture because to considerable extent it controls the ease with which finished wallboard can be dried in the kiln since all of the free water contained in the gypsum board core must be driven through the sheet of paper as water vapor. A sheet which is too dense will be blown off the board in the kiln by the pressure built up beneath it. The inclusion of 2% or more of mineral fiber in the paper sheets produced a highly desirable reduction in the resistance to passage of gas through the sheet.

Water Repellent Paper

This process relates to water repellent cover paper for gypsum sheathing board. Gypsum sheathing board is manufactured by disposing an aqueous slurry of gypsum plaster between two cover sheets while such components are continuously progressing along a conveyor of considerable length. The cover sheets are of a relatively heavy paper, in the order of 0.020" thick, weighing from 60 to 80 pounds per thousand square feet.

N.C. Jacks and J.C. Sparks; U.S. Patent 3,607,486; September 21, 1971; assigned to National Gypsum Co. provide a method of applying a water repellent coating of bituminous material to one surface of a relatively heavy paper sheet with the opposite surface of the sheet remaining substantially free of the coating. It is another object to provide such a method wherein the water repellent coating is applied to the paper sheet after it has been formed and substantially all of the water removed therefrom.

Bituminous materials include native bitumens such as asphalt and the like as well as pyrogenous asphalts, tars and pitches. The water repellent coating composition is formed by dispersing an organic solvent in an aqueous emulsion of bituminous material and the application of this dispersion to one surface of a paper sheet at the dry end of the paper manufacturing process. The organic solvent and aqueous emulsion of bituminous material are blended together in a manner which allows substantially all of the former to be dispersed in the water phase of the emulsion where it acts as a lubricant to substantially prevent the bituminous material from adhering to the apparatus which is employed to apply the water repellent material to the paper. This method may be incorporated conveniently into the standard papermaking process by applying the water repellent coating to the paper as it is being passed through a calender stack. It has been found to be capable of highly efficient operation at calender speeds normally utilized in the papermaking process while at the same time imparting excellent water repellent properties to the paper.

The aqueous emulsion of bituminous material which is used to form the water repellent coating composition may contain up to about 65% by weight of bituminous material. It is preferred to use an emulsified asphalt which is identified as SS-2 throughout the petroleum industry. SS-2, an aqueous asphalt emulsion commercially available from Hunt Oil Company, contains from about 56% to about 64% by weight of asphalt and about 1% of emulsifying agents.

Preferably the water repellent coating composition is formed by lightly blending 7 parts by volume of SS-2 with 2 parts of the organic solvent. The process of blending these two components together must be carried out with low speed, low shear agitation at ambient temperature. For example, the two components may be brought together and blended into the desired dispersion by bubbling air slowly throughout the mixture.

The method of agitation and the temperature conditions are critical for it is essential to the proper functioning of the water repellent coating method that substantially all of the organic solvent be taken into the water phase of the emulsion and not into the asphalt phase. Should any significant part of the organic solvent content of the dispersion be taken into the asphalt phase of the emulsion the resultant dispersion will be tacky and not suitable for use according to the method.

As the severity of the agitation increases and/or the temperature of the dispersion rises to any significant degree the likelihood that the organic solvent may be taken into the asphalt phase of the emulsion becomes progressively greater. The specific temperature at which this condition will occur cannot be precisely defined since its occurrence appears to be dependent upon a time-temperature relationship as well as upon temperature alone.

The dispersion may be tested to determine its suitability for use by centrifuging a sample at high speed for about 30 minutes. A lightly blended dispersion will exhibit a relatively thick layer of organic solvent lying at the surface after undergoing centrifugation whereas if any significant part of the organic solvent content has passed into the asphalt phase of the emulsion the centrifuged sample will have a thick scum on the surface or a relatively thin layer of solvent lying at the surface with a thick scum immediately below it.

Prior to being applied to the surface of the paper the asphalt emulsion-organic solvent dispersion may be further diluted with water if so desired, the degree of dilution varying according to the level of water repellency required for the paper being treated. Where the paper to be treated is gypsum sheathing board cover paper, the dispersion is preferably formed by blending 7 parts by volume of

SS-2 with 2 parts of organic solvent and then further diluting the dispersion with water in a volume ratio of dispersion to water 9:4. The water repellent coating composition thus formed contains from about 30% to about 45% by weight of asphalt. It has been found that excellent water repellent properties are imparted to gypsum sheathing board cover paper when about 100 pounds of the dispersion containing from about 30% to about 45% by weight of asphalt are applied to 2,000 pounds of paper.

The water repellent coating may be applied to the paper by various conventionally known techniques. Preferably the water repellent coating is applied to the paper from a water box on a calender as the paper is passing through the calender stack in the paper manufacturing plant.

The water repellent material is quite easily applied to the fibrous material in the top liner of the paper web in this manner. The coating rapidly loses substantially all of its water and organic solvent content leaving a highly water repellent coating disposed on the fibers of the top liner within a short time interval prior to being collected on the roll. The coating formed is of a relatively viscous material which was applied while in the emulsion state, thus, there is substantially no subsequent penetration of the bituminous material through the paper web to the opposite face nor is there any subsequent transfer of this material to the bottom liner of the adjacent convolution in the ultimate paper roll which will appreciably affect the absorption qualities of the bottom liner. Therefore the bottom liner retains its desired water absorbent properties and the paper sheet has excellent vapor permeable characteristics.

The degree of penetration of the water repellent coating into the paper sheet may be controlled by the sizing present in the paper and the nip pressures of the wet and dry calender stacks. It has been found that highly satisfactory results are obtained when the paper has been hard sized with alum and rosin and the calenders are operated at the nip pressures normally used for the paper being treated. The amount of coating material taken in the paper may be controlled by the percentage of bituminous material present in the dispersion as well as by the type and amount of sizing present in the paper and the nip pressures of the calender stacks.

No. 2-D diesel fuel is preferred as the organic solvent to be blended with the aqueous emulsion of bituminous material to form the dispersion which is ultimately applied to the paper. No. 2-D diesel fuel has a boiling range of from about 335° to 670°F.

Example (A): A dispersion containing from about 30% to about 45% asphalt by weight was applied to the top liner of the paper from a water box on a calender stack in a ratio of approximately 100 pounds of dispersion for each 2,000 pounds of paper.

Example (B): The water absorption of the coated surface of gypsum sheathing board cover paper treated by this method, as determined by TAPPI Test Method T-441 OS-63, when tested for 3 minutes at 120°F was an average of 0.25 gram which is considered to be a highly satisfactory result. When the coated surface of the treated paper was tested according to TAPPI Test Method RC-212, a sliding drop method for determining water repellency, values of from 4 to 5 with an average value of 4.5 were obtained. A value of 4 or above is generally considered throughout the industry to be excellent for gypsum sheathing board cover paper.

Example (C): Identical tests conducted immediately after the coating was applied to the paper showed significantly lesser water repellent properties for the paper. The difference in results obtained at these two different times is thought to occur because substantially all of the remaining organic solvent has an opportunity to escape from the paper upon standing. Thus, tests conducted about 24 hours after the paper has been treated will be indicative of the degree of water repellency impaired to the paper.

PAPERBOARD

Porous Paperboard Containing Plastic Microspheres

According to *H.M. Hoogsteen; U.S. Patent 3,533,908; October 13, 1970; assigned to Brown Co.* a low density high porosity paperboard sheet is provided by incorporating plastic microspheres into the pulp furnish used to form the paperboard. After the paperboard has been formed and dried, it is treated with an organic liquid which is a solvent for the plastic material of which the microspheres are formed. Upon further drying, the paperboard sheet attains a low density and greatly increased porosity, without adversely affecting its other mechanical properties. If desired, the treated sheet may be further subjected to heat treatment at elevated temperature in order to enhance certain physical properties.

Example 1: Preparation of Low Density Fourdrinier Paperboard – A plastic microsphere slurry having the following composition was first prepared utilizing methyl methacrylate microspheres in a Horn Broke beater: pounds of microspheres, 236; gallons of water, 1,370; pounds of water, 11,800; and percent microspheres, 2.0 by weight.

A pulp slurry was then prepared according to the following specifications: pulp, 100% bleached Espanola jack pine; size, 2%; alum, to pH of 4.5 to 5.0; and freeness, 300 to paper machine.

At start-up the paper sheet was formed without the addition of microspheres in order that the heavy sheet could be taken over the machine at maximum strength. After a sheet having the correct weight was obtained (65 pounds), the slurry containing microspheres was added at the fan pump in an amount of approximately 5% microspheres based on oven dry fiber. The sheet has a caliper of about 13 mils. The formed paper sheet was dried in conventional manner and its physical properties measured to give the values shown below. The sheet was then immersed in toluene for 90 minutes maintained at 100 to 200°F. The sheet was dried, its physical properties measured, and further conditioned by heating for 5 minutes in a 325°F oven.

In the table are listed data obtained from tests of the paperboard produced in Example 1 before treatment, after solvent treatment, and after heat treatment. Additionally, the table contains data obtained in experiments listed as Examples 2 through 8 where the process described in Example 1 was repeated, but where the type of pulp, thickness of the paperboard, and percentage by weight of plastic microspheres were varied. In each case, the porosity is given in terms of Sheffield porosity values. Solvent treatment in each case comprised a 90 minute immersion in toluene at 180 to 200°F, with the samples being subsequently dried and reconditioned. In each case, where indicated, subsequent heat treatment comprised placing the sample for five minutes in an oven maintained at 325°F.

The plastic microspheres used may be prepared from any suitable polymeric material such as methyl methacrylate, methyl methacrylate-acrylonitrile copolymers, styrene-acrylonitrile copolymers, phenolic resins, etc. Methyl methacrylate microspheres are preferred. The microspheres generally range in size from about 3 to about 130 microns and have an average apparent density of approximately 0.82 pounds per cubic foot.

The amount of plastic microspheres in the paper sheet may range from a very small amount to as much as 12% by weight. If the amount is too small, the porosity of the paper sheet is not increased sufficiently. On the other hand, amounts greater than about 12% may result in a deterioration in some of the physical properties of the paper sheet. Amounts of from about 1% to about 6% by weight are preferred.

Solvent and Heat Treatment of Microsphere-Containing Fourdrinier Board

Example	Pulp type	Percent by weight microspheres	Before treatment		After solvent treatment		After heat treatment	
			Caliper, inch	Porosity, Sheffield	Caliper, inch	Porosity, Sheffield	Caliper, inch	Porosity, Sheffield
1	Espanola Jack Pine	5.3	.0138	604	.0125	1,400	.0125	1,160
2	do	4.0	.0088	379	.0088	720	.0085	500
3	do	0 [1]	.0079	169	.0078	185	.008	165
4	Bleached kraft	2.0	.020	600	.0193	1,200		
5	do	0 [1]	.0118	540	.0115	580		
6	Espanola Jack Pine	12.0	.035	962	.0318	1,800	.0325	1,530
7	do	6.0	.0205	605	.020	1,240	.0205	960
8	do	0 [1]	.0118	472	.012	550	.012	503

[1] Control.

Although toluene is the solvent utilized in the examples described above, any organic solvent, such as benzene, xylene, acetone, etc., which is a solvent for the particular plastic material utilized to form the microspheres, may be used. Although immersion in the solvent is the preferred procedure for dissolving a portion of the plastic microspheres, other equivalent procedures capable of effecting the same result, i.e., spraying, submergingly translating, or the like may also be used.

The low density high porosity paper or paperboard may be used for many applications. It may be used to produce controlled porosity filter paper, the porosity being controllable by varying the content and size of the microspheres and the degree of treatment in the organic solvent and subsequent drying. The paper or paperboard may also be used to fabricate products requiring a relatively high degree of heat insulation, as for example paper cups to be used for hot beverages. Additionally the paper or paperboard sheets may be used for applications such as the fabrication of battery separators for various types of batteries.

Paperboard of Improved Oil Resistance

J.H. Wing, J.N. Stone and F.B. Thomas III; U.S. Patent 3,953,283; April 27, 1976; assigned to Continental Can Co., Inc. describe preparation of paperboard having high oil resistance wherein a mixture of a cationically active material, a fatty acid salt and a polyfluoroalkyl compound is incorporated in the paper pulp prior to the formation of the pulp into paperboard.

The cationic materials used are cationically modified starches or gums and hydrogen-bonding starches and gums. A wide diversity of cationically active materials may be found in U.S. Patent 3,392,085. Many cationically modified starches and gums are commercially available. A cationic cornstarch is sold by National Starch and Chemical Corp. under the trademark Cato 8.

The fatty acid salt is added to the paper pulp as a soluble metal salt which is insolubilized on the paper fibers using a precipitating agent. The fatty acid salts are generally added to the pulp as either the sodium, potassium or ammonium salts.

The cations of the insolubilized fatty acid salts are aluminum (Al^{+++}), ferric (Fe^{+++}) and/or chromic ions (Cr^{+++}). The fatty acid employed may be any one or a mixture of the saturated fatty acids having from 12 to 20 carbon atoms. For best results, the amount of unsaturated fatty acid component should not exceed 30% by weight of the fatty acid and preferably the fatty acid should not contain greater than 7% unsaturated fatty acids.

The precipitating agents which are employed for insolubilizing the water-soluble saturated fatty acid salts are water-soluble inorganic metal salts of polyvalent metallic ions such as aluminum, ferric and chromic.

One class of water-soluble polyfluoroalkyl compounds which has been found effective is the water-soluble polyfluoroalkyl sulfonamino phosphates. One group of polyfluoroalkyl sulfonamino phosphates which are commercially available are the bisperfluorosulfonamino alkyl phosphates having the general formula:

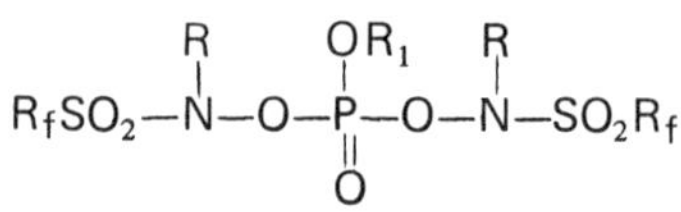

where R_f is a perfluoroalkyl having 4 to 12 carbon atoms, R is hydrogen or an alkyl group having 1 to 6 carbon atoms and R_1 is an alkyl group having 1 to 6 carbon atoms. An example of a commercially available bis-perfluorosulfonamino alkyl phosphate is bis-N-ethyl-2-perfluoroalkyl sulfonamino ethyl phosphate sold by the 3-M Company under the designation FC-807 Paper Chemical.

The cationically active material is added to the pulp suspension at a concentration of 0.10 to 6.0% (preferably 0.70 to 2%) based on the weight of the solids content of the suspension. The water-soluble fatty acid salt is added to the pulp suspension at a concentration of 0.10 to 10.0% (preferably 0.50 to 6.5%) based on the weight of the solids content of the pulp suspension. The polyfluoroalkyl compound is added to the pulp suspension at a concentration of about 0.07 to 0.50% (preferably 0.10 to 0.30%) by weight based on the weight of the solids content of the suspension.

The pulp suspension may be made up with refined fibers such as bleached and unbleached fibers, ground wood, soda pulp fibers, semichemical fibers, kraft fibers, sulfite fibers, textile and synthetic fibers (such as viscose rayon and cellulose acetate), and other cellulosic fibers. The pulp is generally refined to a Canadian Freeness value of from 400 to 700 ml before addition of the cationically active material, fatty acid salt and polyfluoroalkyl compound. The pulp consistency is normally 0.3 to 0.6% by weight, although considerable variation is possible.

Pulp suspensions after refining, normally have pHs of approximately 7.0 to 8.0 It is critical that the pH of the pulp suspension be adjusted to a pH range of from 5.8 to 7.3 and preferably 6.0 to 6.8 during the addition of the cationically active material, fatty acid salt and polyfluoroalkyl compound to the pulp suspension. If the pH of the pulp suspension is outside this critical range, the paperboard product prepared from pulp of this process will exhibit inferior oil resistance.

Example: In this example, the operation of the paper machine to manufacture highly oil resistant paperboard was similar to the method of manufacturing a routine grade paperboard with the exception that a mixture of a cationically active material, a fatty acid salt and a polyfluoroalkyl compound was incorporated in the pulp suspension.

A fiber suspension of bleached pulp containing no additives and consisting of 84% pine and 16% hardwood pulp was refined to a Canadian Standard Freeness of 565 to 578 milliliters and a pulp consistency of 3.5 to 4.1%. The pH of the refined pulp was adjusted to a pH of 8.0 with lime. To the pulp was then added 1.07% by weight (based on the weight of the solids content of the suspension) Cato 8, a cationic corn starch and 4.93% by weight (based on the weight of the solids content of the suspension) potassium stearate and the pH was lowered to varying pHs with 0.5 to 12% by weight (based on the dry weight of the fibers) alum.

This amount of alum was sufficient to completely precipitate the stearate salt on the pulp fibers. Paper chemical FC-807 was added to the pulp suspension as well as 0.4% Vanzak RA-30, a cationic retention aid. The pulp suspension was processed into paperboard in the conventional manner and wound on a reel. Samples were cut from the paperboard roll into square foot sheets and conditioned for 24 hours at 50% relative humidity and 73°F before testing for oil resistance.

The test for oil resistance comprised applying a one half inch head of corn oil (Mazola oil) to the surface of the paperboard and measuring the length of time required for the oil to penetrate the paperboard and stain 50% of the under surface. The results of these tests were recorded in the table below.

For purposes of contrast, paperboard was prepared in accordance with the procedure of the example with the exception that concentrations of FC-807 and pHs outside the scope of the process were used. The results of these contrasting runs are also recorded in the following table and designated by the symbol C.

Test No.	%* FC-807	Pulp pH	Oil Resistance
1	0.10	7.1	20 days
2	0.10	6.2	20 days
C_1	0.05	6.2	6 hours
C_2	0.10	5.5	17 hours

*Based on dry weight of fiber.

MISCELLANEOUS

Self-Bonding Printing Paper

The coated paper of *R.A. Olson; U.S. Patent 3,664,912; May 23, 1972; assigned to P.H. Glatfelter Co.* obtains its unusual printing and self-bonding properties through its combination of coating composition components. In this coating composition, the presence of 20 to 65% of discrete thermoplastic resin particles which remain as discrete particles through the drying operation when the coating is applied to the paper substrate, but which will coalesce under the combined application of sufficient heat and pressure to form a bond to itself or another similar sheet, gives the coated printing paper heat sealing properties generally not found in coated printing papers. These discrete thermoplastic resin particles are used in combination in this coating composition with 15 to 75% of inorganic pigment(s) and 5 to 20% of resinous binder to prepare this coated heat sealing printing paper.

The coated paper obtains its unusual heat and pressure-sensitive properties primarily through the incorporation in the coating composition of 20 to 65%, preferably about 30 to 45%, discrete thermoplastic resin particles, which have coalescing temperatures of at least about 140°F.

Any thermoplastic particles can be utilized in the coating composition as long as they remain as discrete particles and do not act as film-forming binder materials below about 140°F but coalesce at temperatures above about 140°F, preferably in the range of about 140°F to about 400°F. Included are the grades of polyvinyl chloride, polystyrene, styrene-butadiene (80 to 99% styrene to 1 to 20% butadiene), vinyl chloride-vinylidene chloride copolymer, vinyl chloride-vinyl acetate copolymer, terpolymer of styrene, butadiene and acrylonitrile, gum rosin, polymerized rosin, polymerized rosin glycerol ester, partially dimerized rosin, acrylic polymers and powdered polyethylene, having these characteristics.

The thermoplastic particles remain as discrete particles during the drying of the applied coating, so long as the sheet temperature is not raised above the coalescing temperature of the particles. The average particles size preferably ranges from about 0.2 to about 5.0 microns.

In order to give the coated paper the desired properties of proper ink receptivity, ink holdout, ink drying and setting characteristics, permanent opacity (especially at those areas which are heated and pressurized to form the bond), the coating composition contains about 15 to 75%, preferably about 40 to 60%, of one or more inorganic or organic pigments. The average particle size of the pigment can range from about 0.2 microns to about 5.0 microns. For example, inorganic pigments, such as, coating grade clay, calcium carbonate and TiO_2 having average particles sizes of 1.0, 0.20, and 0.25 microns, respectively, have been used as components.

The thermoplastic resin particles do not have sufficient binding or adhesive power (before activation by pressure and heat) to be utilized without the addition of an additional binding or adhesive material to secure the resin and pigment together in the coating layer and to the paper substrate. About 5 to 20%, preferably about 10 to 15% organic resinous binder is used in the coating composition.

Resinous binders which have been utilized either singly or in combination with one another include polyvinyl alcohol, styrene-butadiene copolymer (60% styrene-40% butadiene), soy protein, and hydroxyethylated starch.

Inasmuch as the content of the pigment particles and the thermoplastic resin particles affects the microporosity of the resulting coated paper, the total content of these particles should be in the range of about 80 to 95%, preferably about 85 to about 90%, of the coating on the dry weight basis.

The aqueous coating composition described can be applied to both sides of a white opaque paper substrate, basis weight range between 32 and 100 lb (25 x 38–500), by any conventional coating method, such as blade coating at coat weight levels from about 3 to about 15 lb per side per ream and subsequently dried at a temperature below the coalescing temperature of the thermoplastic particles by means of hot air or steam heated dryer cans or any other means which removes the water or other inert vehicle from the coating so that the final coated sheet has been 2 and 7% final moisture based on the dry weight of paper. The resulting coated paper product has a basis weight range from 40 to 120 lb (25 x 38–500) and has all the appearances and printing properties of conventional clay-coated printing papers.

In addition, the resulting coated paper permits a strong durable bond to be formed between sheets thereof brought into intimate contact under the application of heat and pressure and consequently permits books, pamphlets, brochures, and similar multiple paper forms to be readily produced.

In preparing the coating composition, two basic procedures can be followed. With natural resinous binders and some synthetic adhesives, it is necessary to solubilize the material through a heat or chemical treatment or a combination of both in order to obtain the binding properties of the material upon drying of the coating. With the first procedure, if a resinous binder of this nature is to be used, the material in the dry powder form is blended into an aqueous dispersion of a certain solids content of the pigment particles. The mixture of pigment particles, unsolubilized resinous binder, and water is then subjected to the necessary heat and/or chemical treatment, following procedures outlined by the manufacturer of the resinous binder, to achieve solubilization.

After cooling the mixture to a temperature below the coalescing temperature of the thermoplastic resin, the thermoplastic resin particles, in aqueous dispersion form as received from the manufacturer, are blended into the above described mixture. If an additional resinous binder of the type that requires no solubilization to achieve binding properties is to be used, it is blended into the coating composition next. When additives to improve the characteristics of the coating composition are used, they are blended into the mixture last, usually in the sequence of flow modifiers first and insolubilizers last.

With the second basic procedure for preparing the coating composition, if a natural or synthetic resinous binder requiring solubilization is used, it is prepared by itself, independent of the pigment dispersion, by dispersing the binder in the dry powder form in a certain specified quantity of water and then subjecting the blend to the necessary heat and/or chemical treatment to acheive solubilization. The aqueous solubilized form of the resinous binder is then added to an appropriate amount of inorganic pigment dispersion and the additional components of the coating composition are then added in the sequence described above for the first basic procedure for coating make-up. If desired, other inert vehicles may be used in place of water, although water is by far preferred because of economy, ease of handling and availability.

There are many acceptable methods of heat sealing which can be used with the coated paper. The four major factors which primarily influence the strength of the bonded area are: (1) the power input which is responsible for heating the coating to the coalescing temperatures of the thermoplastic resin particles which varies for the different chemical types of thermoplastic resin used from about 140°F to about 400°F, for example, (2) the pressure applied during heating which varies with thermoplastic resin type and amount of thermoplastic resin in the coating used in this process from about 30 psi to about 2,500 psi (3) the unit of time over which the heat and pressure were applied which varies with type of thermoplastic resin and the amount of thermoplastic resin as well as the heat input and pressure applied from as little as about 0.25 second to several seconds, e.g., 60 seconds, and (4) the width of the bonded area formed which can be varied from about one-sixteenth inch or less to about ¼ inch or more.

Excellent seals have been made with the coated paper depending on the combination of the above mentioned variables using either impulse type heating or radio frequency (RF) type heating devices.

In the examples which follow, unless otherwise designated, all temperatures are in °F, all parts and percentages are by weight and all the percentages listed for coating components are based on dry weight percentages of the total dry solids content of coating compositions.

Example 1: A coating composition comprising the ingredients listed below was prepared by the previously described first basic procedure where the resinous binder is cooked in the presence of pigment and water sufficient to adjust the solids content to 50.6% solids.

	Weight Percent
Thermoplastic resin– Polyvinyl chloride as an aqueous latex, Geon 151, containing about 56% solids polyvinyl chloride, having a coalescing temperature above 250°F and an average particle size of about 0.2 micron	30.0
Inorganic pigments–	
(1) Clay pigment available as Ultra White 90 having 92 to 94% particles of less than 2.0 microns	55.6
(2) Calcium carbonate pigment available as Purecal-O having a particle size of 0.10 to 0.35 micron	6.2
Resinous binder–Polyvinyl aclohol available as Elvanol 71-30, having a molecular weight of 133,000, 99 to 100% vinyl alcohol units and 0 to 1% vinyl acetate units	7.0

(continued)

	Weight Percent
Flow modifier–Calcium stearate as an aqueous dispersion sold as Nopcote C-104 and containing 50% calcium stearate	1.0
Binder insolubilizer–Glyoxal as an aqueous solution containing 40% glyoxal	0.25
Total:	100.05

The composition was applied to both sides of 3.0 mil thick white base paper at 10 to 13 lb per ream (total of both sides) using a trailing blade coater and the coated paper was dried with forced hot air at 220°F for 5 seconds per side. The coated paper was rewound and sheeted. Some of the paper was supercalendered to enhance its gloss to a value comparable to other commercial glossy coated printing papers.

Example 2: A coating composition comprising the ingredients listed below was prepared to 46% solids as described in Example 1.

	Weight Percent
Thermoplastic resin–Styrene-butadiene copolymer as a 48% aqueous dispersion, consisting of 80 parts polymerized styrene and 20 parts polymerized butadiene, having an average particle size of about 0.2 micron and a coalescing temperature of about 140°F, sold as SD-345	62.0
Inorganic pigment–Calcium carbonate pigment as described in Example 1	33.0
Resinous binder–Polyvinyl alcohol as described in Example 1	5.0
Total:	100.0

The composition was applied to both sides of 3.0 mil thick white base paper at a total of 8 to 10 lb per ream (total of both sides) using a trailing blade coater. The coated paper was dried with forced hot air at 220°F for 5 seconds per side. The table below represents a comparison of the papers of this process with conventional coated printing papers and with a selected plastic "paper," and a plastic-coated paper. Coated papers of the process were those made in Example 1 (unsupercalendered paper), and Example 2. Papers A through D represent typical commercial coated printing papers. Paper A1 is a sheet of typical paper that has bound on it a layer of extruded polyethylene. Paper B1 is an "imitation paper," made entirely from a plastic material and known as Ucar.

	Papers of -- Process --		----- Conventional Coated Printing Papers -----				Plastic Coated or Plastic Paper	
Tests	Ex. 1	Ex. 2	A	B	C	D	A1	B1
GE brightness	83.7	76.9	77.8	82.3	81.1	80.5	74.6	86.8
Opacity	93.0	88.0	92.1	93.6	93.5	95.0	91.5	88.2
Gloss	18.5	12.0	13.0	8.0	15.0	58.0	55.0	35.0
K and N ink receptivity (% brightness retained)	57.3	75.8	58.0	57.0	61.8	76.3	98.2	91.7
Ink set-off (% transmission)	97.7	71.5	89.5	80.1	92.9	80.0	24.8	41.2
Heat seal strength RF sealing (lb/in)	0.46	0.62	None	None	None	None	0.82	None
Heat seal strength impulse sealing (lb/in)	0.68	0.58	None	None	None	None	1.34	1.64

The G.E. Brightness, Opacity, and Gloss tests were run according to TAPPI Standard Test Methods T-452, T-425, and T-480 respectively. The K & N Ink Receptivity Test was performed similar to TAPPI Routine Control Method RC-19, except that the dyed test ink was left on the paper being tested for only one minute instead of the prescribed two minutes. After the test was completed, a G.E. Brightness value was obtained on the tested area of the paper and this value was divided by the original brightness of the paper and multiplied by 100, yielding a value known as the percent brightness retained by the test area. The lower this value, the more receptive the paper is to the ink.

Coated printing papers must achieve a balance between ink receptive and ink holdout properties. Too much ink holdout could result in ink drying problems that might cause set-off of ink to sheets placed on top of the printed sheet, or even more serious, the sheet may not even accept a normal printing ink. On the other hand, coated printing papers must exhibit a certain amount of ink holdout for glossy prints and good definition.

To further compare these papers for ink receptivity and ink set-off characteristics, an ink set-off test was performed using the IGT Printability Test in conjunction with the Westvaco 20 mm Printing Disc. The ink used was IPI Speed King Process Blue Litho Ink. Generally, this procedure consists of metering ink onto the Westvaco Disc, transferring this ink at 50 kg pressure and M speed setting using the IGT Printability Tester to the sample being tested, and then after a specified time (samples in the above table were tested after 30 minutes), placing a sheet of the same paper on top of the printed paper, clamping both samples in the tester, and putting an impression on both samples using the IGT 20 mm Printing Disc at a pressure of 50 kg. This impression pressure causes undried ink on the original printed sample to set-off to the superimposed sample.

The degree of ink set-off was measured by using a Densichron, manufactured by W. M. Welch Scientific Company, Chicago, Ill., with the red filter and obtaining a percent transmission reading on the set-off area first by setting the instrument to read 100% transmission on the unprinted paper and then obtaining the reading of the ink set-off area. Therefore, the higher this value, the more light is reflected from the paper surface and the less set-off of ink has occurred.

Heat seals were made with both a Radio Frequency (RF) and impulse heat sealer. With the Radio Frequency Sealer, pressure of 2,300 psi and heat at a temperature range of 300° to 350°F were applied over an area 4 inches by 1/16 inch for 0.25 second. With the impulse sealer, pressure of 233 psi and heat at a temperature of 300°F were applied over an area 1 inch by ½ inch for 4 seconds.

Heat seal strength was measured by employing a peel-type tensile test on the sealed sheets. For sheets sealed by RF, a 3-inch width seal was measured; for sheets sealed with the impulse sealer, a 1-inch width seal was tested. The test was conducted by clamping the ends furthest from the seal of the respective bonded sheets in special jaws of an Instron Tensile Tester and the force required to peel the sheets apart was measured in pounds per inch.

From the results of the table, it is apparent that the papers of the process are comparable to the conventional coated printing papers in optical and ink receptive properties. The plastic coated and plastic "paper" vary in optical properties but could compare to conventional coated papers. However, the ink receptive and set-off characteristics of these papers is so poor that normal printing inks could not be used to print them conventionally. The seal strengths of the papers of this process are not as great as in the plastic or plastic-coated papers. However, the seals of the papers of the process are adequate for binding by nature of the seals' fiber pulling ability when peeled apart.

Single Continuous Operation for Manufacture of Planographic Printing Paper

D.A. Newman; U.S. Patent 3,592,730; July 13, 1971; assigned to Columbia Ribbon and Carbon Manufacturing Co. describes a process for making planographic printing paper in a continuous operation on a papermaking machine. A paper web is formed from an aqueous slurry and dried to the point that it contains from about 5% to about 30% moisture and is self-supporting. A coating of insolubilizing sizing composition, comprising a hydrophilic film-forming binder material such as polyvinyl alcohol is applied to both sides of the web, the sized web is heated sufficiently to dry the sizing composition to the point that it is non-tacky, but without substantial insolubilization of the size, and then a coating of self-insolubilizing planographic composition, comprising 1 part by weight of a hydrophilic film-forming binder material, such as polyvinyl alcohol, an insolubilizing agent such as glyoxal, and from 4 to 8 parts by weight of a filler, such as clay, silica, or a mixture of both, is applied to at least one side of the web. Finally the coatings are insolubilized together and the web is calendered, all on the papermaking machine.

Example: A paper furnish is prepared in conventional manner from beaten pulp in amounts sufficient to form a final web weighing about 20 pounds per ream of 500 sheets, 25" x 38" in dimension. The furnish may have melamine-formaldehyde resin added thereto in order to impart wet strength to the web. The furnish is processed through a papermaking machine in which the pulp is collected in sheet form on a screen and the water is filtered away. The sheet is contacted with drying cylinders and heated sufficiently to reduce the moisture content to about 10% by weight and produce a compacted self-supporting paper web.

The moist web is next passed in the nip of coating rollers which apply about 3 pounds per ream of sizing composition to each side of the web, the composition comprising the following:

Ingredients	Parts by weight
Polyvinyl alcohol (10% aqueous solution)	10.0
Glyoxal (40% aqueous solution)	0.25
Clay (60% aqueous dispersion)	4.0
Syton (colloidal silica)	3.0

The coated web is next contacted with drying drums heated to about 300°F to dry the size coatings to the point that they are not adhesive and still retain some moisture. The duration of heating is brief so that the insolubilization reaction is not permitted to proceed very far but may be initiated.

Next the size-coated web proceeds into the nip of a second set of coating rollers which apply about 3 pounds per ream of planographic composition to each surface of the web, the composition being as follows.

Ingredients	Parts by weight
Polyvinyl alcohol (10% aqueous solution)	10.0
Clay (60% aqueous suspension)	8.0
Syton (colloidal silica)	3.0
Glyoxal (40% aqueous solution)	0.5

The coated web is then passed against a series of drying drums heated to conventional papermaking temperatures to dry the web, size coatings and planographic coatings to leave about 5% moisture content and cause the coatings to insolubilize and harden together as an integrated unit which in turn is integrated with the paper web.

Finally the web is contacted with calendering rolls which smooth the planographic surfaces and compress the web, and the web is wound on a take-up roll for final cutting. This winding operation is necessary because of the speed of operation of the papermaking machine. The cutting of the web into sheet lengths is preferably done as soon after winding as possible.

While it is not essential that the present plates carry a planographic printing layer on both surfaces, it is essential that they carry a self-insolubilizable curl-compensating layer on the surface opposite to that carrying the planographic printing layer.

Preferably, the curl-compensating layer is a planographic printing layer of the same composition as the printing layer on the opposite surface so that either or both surfaces may be used for printing purposes. Otherwise, the curl compensating layer should be tinted for identification purposes and may comprise a coating of a water-applied self-insolubilizable binder material. The binder material is preferably a hydrophilic material such as polyvinyl alcohol, carboxymethylcellulose, starch, casein or the like, in combination with a suitable hardening agent.

The hydrophilic nature of the sizing composition applied to both sides of the paper web and integrated therewith improves the printing properties of the present plates by reinforcing the hydrophilic properties of the top planographic layer and masking the fibers of the paper support. Equally important, the sizing layers permit the making of erasures and corrections on the planographic layer since the removal of portions of the planographic layer during the erasing operation will merely expose portions of the sizing layer which are also hydrophilic and planographic.

Catalysts to Promote Chemical Drying of Ink Incorporated into Paper or Coating

P.G. Bleier; U.S. Patent 3,711,317; January 16, 1973; assigned to Samum Vereinigte Papier-Industrie, K.G. Austria is concerned with improving the printability of printable material. It has been ascertained that the beneficial effect of catalysts in promoting the chemical drying of printing inks is not impeded by the presence of mineral materials, such as clay, provided the catalysts are available on the surface of the material to be printed in finely dispersed form, as distinguished from the usual continuous film.

The catalysts may be incorporated into clay containing coating masses which are used for coating printable material such as papers. The presence of the catalysts in finely dispersed form causes a more rapid and effective drying of the printing ink and results in an improved solidification of the printing ink film on the printed material.

For uncoated papers the catalytic substances are either admixed with the fibrous paper material during the paper manufacture or are added to the surface size which is used in the paper manufacture. Thus, for example, the substances may be added to the machine vat before the paper mass enters the papermaking machine or they may be incorporated into the liquid formulation that is applied to the paper at the size press of the paper machine. The catalysts become fixed on the paper fibers and no extraneous surface coating prior to printing is necessary.

Essentially, therefore, the basic thought underlying this process is to incorporate catalytic substances, which have a tendency to promote or facilitate crosslinking of printing ink components and which, generally were contained in a separate cover film, directly into the paper or coating mass in finely dispersed form.

Particularly suitable catalysts are metal salts of higher organic acids wherein the metal is cobalt, manganese or zirconium. The acid moiety of the salt may be octoic acid or naphthenic acid. It has also been found that excellent results are obtained if the catalysts are substances which release active oxygen. Such substances are, for example, normally stable, inorganic or organic peroxides, such as barium peroxide, zinc peroxide, sodium perborate, dilauroyl peroxide and dibenzoyl peroxide.

Excellent results are obtained if the printable material contains the following quantities of catalyst: (a) if the catalyst is an organic metal salt (expressed as ppm of the active metal): about 1 to 100 ppm; (b) if the catalyst is an inorganic peroxide: about 0.1 to 5% by weight; and (c) if the catalyst is an organic peroxide: about 1 to 100 ppm. Although the process is applicable to both coated and uncoated materials, the primary use is with coated materials such as pigment coated paper, since printing processes are primarily carried out on the latter type of material.

In accordance with one modification the catalysts are incorporated into the pigmented coating in the form of a dispersion forming an extraneous phase. For this purpose the catalysts are added in the form of an oily dispersion to the aqueous pigment coating mass or mix. The catalyst particles are finely distributed in the aqueous pigment coating mass by means of surface activity which may be enhanced by the addition of surface active substances. Surface active or wetting agents of the following type have proved to be suitable: (a) nonionic wetting agents: condensation products with ethylene oxide of different chain length; and (b) anionic wetting agents: sulfonates like dodecylbenzenesulfonate, sulfonated castor oil, sulfonated olefinic hydrocarbons.

This dispersion, while maintaining its catalytic activity, is stabilized by means of protective colloids and fixed in the forming capillary system during the drying of the coating on the paper base. The activation of the catalysts is delayed and is triggered by the contact or wetting with the oily printing ink components, whereupon the catalysts are released from their local fixation. Once the catalytic activity has been initiated, the catalysts exert a strong drying action, particularly due to their fine distribution, and this catalytic action is very rapid and effective and promotes the drying of the drying oils and resins of the printing ink.

Example: For producing high grade coated art papers with built-in drying acceleration, to be used with letter press and offset printing inks, the drying promoting agent or catalyst is added to the prepared coating mass or mix for the surface application in the form of cobalt octoate, the cobalt salt of 2-ethylhexanoic acid.

Base paper free of mechanical wood fibers and weighing 75 g/m^2 is coated with an aqueous coating mix of 40 to 50% solid ingredients on both sides to a final dry weight of 100 g/m^2 on a smoothing roll coating machine. The coating mix consists, as far as pigments are concerned, of about equal parts of calcium carbonate, satin white, and English china clay, and as to binders of equal parts of sodium caseinate and a copolymer of acrylate and polyvinyl acetate (trade name Acronal). The coated paper web is then once more passed through a smoothing roller coating machine, where a top coating is applied in an amount of 10 g/m^2 on each side.

The mix used for the top coat is of the same composition as indicated above with the same solid contents; however, cobalt octoate was added to the coating mix in the amount of 0.005% of cobalt, calculated on the solids of the mix. By first diluting the cobalt octoate with the tenfold amount of commercial isopropyl alcohol, a homogenous distribution of the active agent is accomplished. By incorporating the cobalt octoate, which is dissolved in an oily phase, under vigorous stirring into the aqueous coating mix, the former separates into extremely fine particles. Coalescence of the particles is prevented by the binders which are present in the mix and which also act as protective colloids.

In the above case, the active top coating dried completely within 7 seconds; at this drying rate, a migration of the oily droplets into the upper strata of the top coating may be assumed, that is to say, the droplets are spaced not more than a few microns from the surface. The coated paper is calendered on a 16 roller or bowl calender with a linear pressure of 160 kg/cm, and is cut into sheets. The admixture of additional surface active agents was not necessary in this example.

Such sheets were printed in a printing shop at an hourly output of 4,000 to 4,500 sheets with half-tone etching using a screen with 60 dots/linear centimeter; printing was done on both sides in single color (black) with a commercial, pure printing ink as it comes out of the can. For comparison, a commercial high-grade art paper, grade enamel No. 1, having the same substance, was printed in the same manner. The prints made with the papers activated by addition of cobalt octoate did not exhibit offsetting, to wit, rubbing off, and had superior abrasion resistance, and the staple or stack could be folded back immediately. No dusting was necessary.

Contrary thereto, light dusting was required when work was done with the commercial papers. Both rubbing-off time (wiping resistance immediately after printing) and complete drying (nail test after 5 and 24 hours) were found improved. In the test a slow drying, inexpensive job printing ink was intentionally used.

The improved drying characteristics of the catalyst-treated art paper were established as follows: The paper was run in an offset four color sheet printing press without using any dusting. Immediately after a stack of 1,000 sheets had been formed, which took about 12 minutes, the stack was removed from the press. For testing, the printer tried to rub away the drying printing ink film and estimated the pressure of his finger tips required to do so. The stack was then beaten heavily with a heavy wooden club. Then the back side of the sheets was inspected below the spot where the stack had been hit in order to see whether any offsetting had taken place due to the force by which the fresh ink film had been pressed onto the back side of the overlaying sheet.

The first test showed that the cobalt treated printing papers had achieved a state of dryness after about 3 minutes which was reached by the untreated papers after about 15 minutes only.

The second method of testing indicated that no offsetting whatsoever had taken place in the stack containing cobalt treated papers. The back side of the printed papers of the untreated stack showed a clear offsetting of ink and also, on separating the different sheets of the stack, one could hear a sound typical for the separation of sticky materials.

One week after the printing, a comparison test in respect to the abrasion resistance of the prints was carried out. The printed papers, to wit, both the cobalt octoate treated printed paper and the printed commercial paper were subjected to abrasion in a conventional abrasion testing apparatus for three minutes and at a load of 1 kp and a rotation speed of about 500 rpm. The test indicated a considerably higher abrasion resistance of the prints on the cobalt activated paper of the process as compared to the abrasion resistance of the prints on the commercial paper.

Laminate of Expandable Plastic Beads Between Two Cellulose Fiber Webs

H.G.F. Nisser, J. Petersen and S.E. Porrvik; U.S. Patent 3,941,634; March 2, 1976; assigned to Kemanord Aktiebolag, Sweden describe a method for the preparation of a composite that contains plastic particles which includes:

(a) forming a first partially dewatered web of cellulose fibers (dry content 8 to 20% by weight) by depositing the fibers from a pulp on a wire support;

(b) depositing on the first web a plurality of expandable plastic beads, the plastic being selected from copolymers of acrylonitrile and vinylidene chloride, polystyrene or copolymers thereof, or polymethyl methacrylate or copolymers thereof, in a quantity of 0.05 to 60% by weight of the plastic particles based on dry cellulose fibers, the unexpanded beads having diameters of 0.5 to 200 μm;

(c) depositing over the beads a partially dewatered second web of cellulose fibers formed by depositing the fibers from a pulp on a wire support so that the beads are disposed between the first and second webs;

(d) pressing the first and second partially dewatered webs together; and

(e) subjecting the pressed together webs to a temperature of at least 75°C to effect both a drying of the webs and an expansion of at least a portion of the expandable beads.

Plastic beads of polystyrene or copolymers with at least 70 weight percent styrene are economical. The plastic dispersion can also contain cellulose fibers and conventional papermaking additives. The weight ratio of plastic beads to cellulose fibers should be 1:20 to 20:1.

The expandable plastic beads can be introduced between the two wet webs either in dry form or in wet form. If dry application of the polymer is selected the distribution of the plastic beads between the wet webs must be carefully regulated in order to avoid over-dosage with the subsequent lump formation. As examples of suitable application forms can be mentioned pneumatic deposition or screening. With these types of dry application a certain sorting out of coarse and fine fractions of the polymer beads will simultaneously be obtained and thus the demand for uniform particle size of the polymer will be reduced.

With wet application of the polymer beads a dispersion of the beads in water or other inert dispersing medium is suitably used. The particle size range of the thus dispersed polymer should be kept rather narrow to avoid too great variations in the finished product. The dry content of the dispersion is dependent on the amount of plastic beads desired in the finished paper and should normally be within the range of from 0.05 to 40% by weight based on the dispersion.

Example: On a paper machine equipped with two wire screens located in such a way that the formed fiber webs can be brought together before the couch is made on each wire, two webs of sulfate pulp were formed having a weight of 66 g/m^2. On the web on the lower wire was then applied before couching a water dispersion containing 18 g/l of expandable polystyrene beads in an amount equivalent to 65 g polystyrene per square meter of sheet. The dispersion water was suctioned off by a suction box placed below the wire, whereafter the upper web was applied.

After pressing in the press section the webs with intermediate plastic were dried at a temperature of 90°C in the drying section. The obtained sheet had a weight of 197 g/m^2. On comparison with a conventional sheet of the same paper making pulp, but without a polystyrene layer the following results were obtained:

Polystyrene, %	30	0
Basis weight, g/m^2	197	200
Thickness, μm	402	321
Bulk, cm^3/g	2.0	1.6
Z-strength, kPa	410	240
Longitudinal stiffness, 10 mmg	250	204
Transverse stiffness, 10 mmg	165	103

Multi-Ply Paper Having a Wet-Laid Ply and a Dry-Laid Ply

H.G. Curry, B.W. Attwood and D.G.W. White; U.S. Patent 3,954,554; May 4, 1976; assigned to Karl Kroyer St. Anne's Limited, England describe a method and apparatus for making a multi-ply paper sheet wherein a first web is formed by wet-laying fibers and a second web is formed by dry-laying fibers, the two webs being combined to form a multi-ply sheet and wherein the dry-laid web is produced by depositing dry fibers on a supporting band and consolidating the web by hot moist pressing before transferring to the first wet-laid web.

The process is described with reference to Figure 9.4. The apparatus comprises three pieces of basic equipment, namely, a Fourdrinier forming wet end **11**, a device for laying a web of dry fibers **12**, and a cylinder mold machine **13**. The Fourdrinier wet end **11** and cylinder mold are both adapted to form webs of wet laid fiber in conventional manner. Stock from the head-box **14** of the

Fourdrinier wet end **11** deposits fibers onto the forming wire **15** to form a wet-laid web **16** by dewatering through the wire.

FIGURE 9.4: APPARATUS FOR MULTI-PLY PAPER HAVING A WET-LAID PLY AND A DRY-LAID PLY

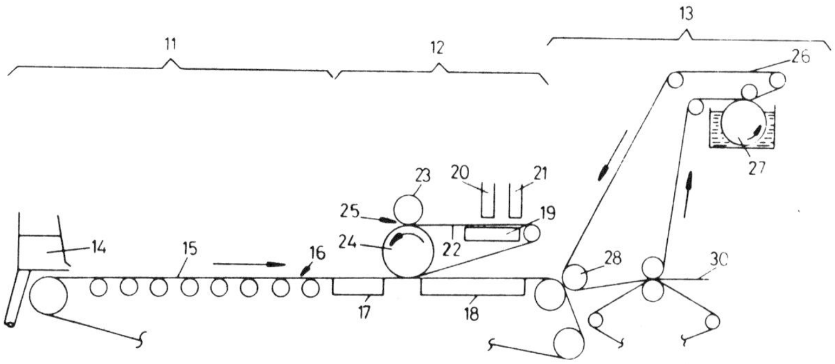

Source: U.S. Patent 3,954,554

Dry fibers mixed with dry powdered adhesive or binder are deposited in an air stream through two heads **20** and **21** onto a foraminous forming wire **22** in the device **12**. A vacuum box **19** assists in drawing the fibers onto the wire **22**. The web of dry fibers so formed is consolidated by a press roll **23** defining with a heated cylinder **24** a pressure nip **25**. The formed and consolidated web is transferred from wire **22** onto the upper surface of the wet-laid web **16** by suction boxes **17** and **18** to produce a two-ply web.

In the cylinder mold machine **13** stock is deposited onto the forming wire **26** by cylinder **27**. The formed web on wire **26** is deposited onto the upper surface of the two-ply product leaving wire **15** at roll **28** to form a three-ply product **30** which passes onto the drying and surface finishing sections of the machine (not shown).

With the apparatus there is provided a means for producing a three-ply product having wet-laid, dry-laid and wet-laid plies. It will be appreciated that further plies can be added by the wet or dry process as required.

The adhesive or binder mixed with the dry fibers in the second ply is activated by the water present in the wet-laid plies to provide a bonding agent for laminating the three plies together. Furthermore there is sufficient moisture in the wet-laid plies to provide some hydrogen bonding between the fibers in the wet and dry plies in the finished product. However, compared with conventional wholly wet-laid sheets and wholly dry-laid sheets of material this combination machine forming wet and dry plies in a single product has considerable advantages since it enables a manufacturer to enjoy the characteristics conventionally provided by a wet-laid product whilst enjoying the additional benefits of the dry-laid web.

A machine of the type shown and having only a Fourdrinier section **11** and cylinder mold section **13** producing a paperboard of 100 g/m^2 can, without altering its speed, have its output doubled to provide a 200 g/m^2 product merely by adding a relatively small dry-laid section **12**. No additional drying capacity is required and no additional plumbing and water catchment devices are necessary. Not only does such an addition substantially increase the output of conventional machines but by suitably modifying the weight of the dry-laid web being added to the machine the output of the machine can be varied considerably. It is possible to alter the weight of web being deposited in the dry-laying device **12** with considerable ease compared with the facility available in conventional wet-laying sections **11** and **13**. It is possible in practice to vary the output of a typical wet end machine from say 100 g/m^2 to 200 g/m^2 at any intervening step by relatively simple adjustment of the capacity of the dry-laying section **12**.

A variation of the apparatus (not shown) comprises a Fourdrinier wet end, a device for forming a dry-laid web and a further Fourdrinier section for forming a second wet-laid web. The last section replaces the cylinder mold device **13**.

Parchmentized Webs—Controlling Temperature of Wash Liquid

Parchmentizing is usually effected by passing the web of raw material through a parchmentizing liquid which usually is sulfuric acid but which may be any of the acid, neutral or alkaline liquids known to parchmentize at suitable times, concentrations, and temperatures. Following parchmentizing the web is washed with successively more dilute aqueous solutions of the substance used for parchmentizing and care must be taken to avoid the development of insufficient strength in the wet web during parchmentizing such as may cause breaks in the web and interrupt production.

P.J. Gilder; U.S. Patent 3,617,200; assigned to Wiggins Teape Research & Development Ltd., England provides a method of preparing a parchmentized fibrous web in which the web is drawn from a parchmentizing bath and is passed to successive washing positions, characterized by the step of maintaining the temperature of the washing liquid applied to the web at the first washing position at a substantially constant level.

It has been found that by controlling the temperature of the washing liquid at the first washing position the properties of the product may be varied. Further advantages are reduced cost by operation at a higher linear speed of the web without loss of product quality due to reduction of material lost at the first washing position, fewer breaks in the web during the process, and increased efficiency in acid recovery due to the formation of less black amorphous scale. (As is well understood, in the recovery plant associated with the parchmentizing process to recover the strong acid the discoloration thereof is removed by an oxidation reaction, for example by the addition thereto of an oxidizing agent such as potassium permanganate or nitric acid, and a further advantage of the process is that less of such an agent is required than is the case with the previously known parchmentizing processes.) There are several variations by which the process can be carried out.

In one, the first washing position is a bath containing the washing liquid; the web is passed through the washing liquid, and the temperature of the liquid is maintained constant by a thermostatically controlled heat exchange means.

In another the first washing position is a bath containing the washing liquid, and washing liquid is extracted from the bath and is recirculated to the bath, and the temperature of the washing liquid is maintained at the substantially constant level by washing liquid recirculated to the bath through thermostatically controlled heat exchange means. In yet another variation the washing liquid may be caused to flow to the bath at the first washing position from a second bath at a second washing position and the temperature of the washing liquid in the bath at the first washing position is maintained at the substantially constant level by thermostatically controlled heat exchange means immersed in the liquid in the second bath.

Alternatively, the washing liquid may be caused to flow to the bath at the first washing position from a second bath at a second washing position and the temperature of the washing liquid in the bath at the first washing position is maintained at the substantially constant level by thermostatically controlled heat exchange means located at a position between the position at which the liquid leaves the second bath and the position at which it enters the bath at the first washing position.

The method may include the step of agitating the washing liquid in the bath at the first washing position to equalize the temperature of the washing liquid in the bath. In another modification the washing liquid is applied to the web by spraying means at the first washing position, the washing liquid is delivered to the spraying means from a collecting trough at a second washing position, and the temperature of the washing liquid applied to the spraying means is maintained at the substantially constant level by thermostatically controlled heat exchange means immersed in the liquid in the collecting trough.

In a still further embodiment the washing liquid is applied to the web by spraying means at the first washing position, the washing liquid is delivered to the spraying means from a collecting trough at a second washing position, and the temperature of the washing liquid applied to the spraying means is maintained at the substantially constant level by thermostatically controlled heat exchange means at a position between the position at which the liquid leaves the collecting trough and the spraying position for the first washing position.

It is found that the greatest advantage is obtained by the first 10°C of cooling and the advantage is greater as the temperature increases. For example, a greater advantage is achieved by cooling from 40° to 30°C than is obtained by cooling from 30° to 20°C. The preferred temperature of the liquid in the first washing bath is +15°C, but it may be as low as -15°C. The benefits increase with the machine speed because this increases the volume entering the first washing position and hence the temperature of the liquid in the first position. The temperatures mentioned above are typical for web speeds of up to 50 to 120 meters per minute but the web speed may be up to 250 meters per minute.

Example: Base paper consisting of 100% bleached sulfate wood pulp containing proportions of virgin pulp, 36% softwood and 64% hardwood with a basis weight of 53 g/m^2 was subjected to parchmentizing with 68.7% sulfuric acid for 5 seconds at 20°C. The concentration of the first washing bath was 33.7% sulfuric acid. By reducing the temperature of the liquid at the first washing bath from 50° to 12°C parchment properties were affected as follows:

	% Change
Reduction in light scattering coefficient	33
Increase in wet tensile strength	22
Increase in dry tensile strength	16

The temperature decrease also reduced the number of pinholes per square meter from 300 to 0 and correspondingly increased the resistance to grease absorption. In a laboratory experiment where shrinkage was controlled, reducing the temperature decreased the cellulose loss to the bath, resulting in an increase in substance of nearly 4%.

Cigarette Paper—Conversion of Cellulosic Fibers by Selective Oxidation

T. Briskin and G.R. Ward; U.S. Patent 3,640,285; February 8, 1972; assigned to Sutton Research Corp. provide a method for producing a paper or wrapper of cellulosic material which is characterized by having improved odor, taste, and aroma during pyrolysis whereby cigars and cigarettes currently produced with tobacco can materially be improved, at least from the standpoint of taste and aroma.

This is achieved when the calcium oxalate and/or magnesium oxalate or corresponding mineralizing agent, is present in the paper in an amount of 5 to 45%, and preferably 10 to 40% by weight. The desired composition can be obtained by the substitution of calcium oxalate and/or magnesium oxalate for calcium carbonate in the finish or slurry of cellulosic pulp used in the original formation of the paper web.

Instead of oxalic acid, use can be made of pivalic acid or tannic acid to form the corresponding pivalate or tannate. Instead of calcium, use can be made of magnesium, barium, or strontium to provide the corresponding salt in the cigarette paper. It is preferred to make use of the combination of a major proportion of calcium and a minor proportion of magnesium as the cation of which the filler is formed.

By way of modification of this initial concept, it has been found that still further improvement in the smoking characteristics of the paper can be obtained when a small amount of free pivalic, tannic and preferably oxalic acid is reintroduced or otherwise provided in the cigarette paper. The function of the free acid returned to the treated paper is to ensure that small amounts of ions of heavy metals, such as calcium, barium, strontium, magnesium, copper, cobalt, iron, and aluminum remaining in the paper or subsequently released are taken up to form the corresponding insoluble heat-stable oxalate, tannate or pivalate, as the case may be.

It is preferred to introduce the oxalic acid in an amount of 0.1 to 0.25% by weight. Reintroduction can be made by dissolving the calculated amount of acid in an aqueous medium to provide a solution containing 0.1 to 1% by weight of the acid and then spraying the solution onto the web or by passing the web through a solution followed by drying, or by mixing the solution with the pulp fiber slurry during the papermaking.

Cigarette paper as it is prepared from purified cellulose loaded with 20 to 30% by weight calcium carbonate is considerably improved by treatment with a dilute solution of oxalic acid containing from 1 to 10% by weight oxalic acid in a dissolved state in an aqueous medium at a temperature within the range of room temperature to 80°C.

The oxalic acid solution operates to convert calcium carbonate to insoluble calcium oxalate. The latter is sufficiently stable thermally to remain while the cellulosic material is pyrolyzed during the smoking of the cigarette so that the effect of calcium on the odor is minimized and a cleaner tasting cigarette is obtained. During the final stages of the pyrolysis, the calcium oxalate breaks down to calcium oxide and calcium carbonate which remain as ash and carbon dioxide and/or carbon monoxide gas which goes off with the smoke and is substantially odorless and tasteless.

In a modification, the prepared cigarette paper, and preferably the purified cellulose pulp of which the cigarette paper is formed, is treated to effect selective oxidation of the methylol groups on the C_6 position of the cellulose molecule with permissible oxidation at the C_2 and/or C_3 position to produce a preferably partially oxidized cellulose hereinafter referred to as oxycellulose in which the oxidation reaction product may be comprised of polyuronic acids and their derivatives.

The amount of oxidation of the cellulosic material should be limited to within the range of 40 to 60% calculated on the conversion of the methylol group on the C_6 position of the cellulose molecule. In effecting the desired level of selective oxidation, it is preferred to make use of liquid nitrogen dioxide as the oxidizing medium.

Treatment can be made with an amount of liquid nitrogen dioxide, preferably 25 to 50 times the cellulosic material on a weight basis. The oxidation reaction should be carried out with the materials at a temperature within the range of 15° to 65°C and under autogenous pressure.

It is desirable to make use of an oxidizing medium of liquid nitrogen dioxide containing 0.5 to 5% by weight water and preferably 1.5 to 3.5% by weight water. At a temperature of 40° to 45°C, an oxidation level of 40 to 60% can be reached in a matter of minutes to hours thereby to permit oxidation as a continuous operation by passage of the paper continuously through the oxidizing medium under autogenous pressure.

The rate and uniformity of oxidation can still further be improved by means of introducing into the reaction medium of gaseous oxygen or a molecular oxygen-containing gas, such as air or oxygen-enriched air, but in which any gas in which the oxygen is included is an inert gas. The oxygen not only serves as a means to agitate or stir the ingredients, but, more importantly, it provides an oxidizing function of its own in the presence of nitrogen dioxide thereby to contribute to the oxidation and the rate of oxidation of the cellulosic fibers. It operates further in the system to oxidize formed nitric oxide to nitrogen dioxide thereby to maintain the level of oxidizing medium in the system. When used, the amount of oxygen used should range from equimolecular proportions with the cellulosic segments making up the cellulosic material (such as glucose or hexose) to about twenty times the theoretical amount.

Further modification, additionally to improve the taste and aroma of cigarette paper and smoking products formed thereof, can be achieved by subsequent treatments of the oxycellulose produced as by subjecting the cellulosic material selectively oxidized with nitrogen dioxide to a reduction reaction with a borohydride or nascent hydrogen.

As the borohydride, it is preferred to make use of an alkali metal borohydride such as the borohydride of sodium, potassium or lithium, dissolved in aqueous medium in an amount within the range of 0.5 to 5% (preferably 0.1 to 1%) by weight. It is desirable to carry out the reaction with materials present in the ratio of one part by weight of the oxidized cellulosic derivative to 10 to 1,000 parts by weight of the solution and preferably 20 to 100 parts by weight of the solution. Results can be secured with a residence time of 1 minute or more but it is preferred to make use of a reaction time within the range of 5 to 60 minutes.

The desired reduction reaction with a minimum of degradation or dissolution can be achieved by adjustment of the solution of the borohydride to a pH of 6 to 7. The reaction can be carried out at a temperature within the range of 0° to 30°C, (preferably 0° to 25°C).

Example 1: Ashing Ingredients Introduced Prior to Paper Formation — Purified cellulose pulp is slurried with a mixture of calcium oxalate and magnesium oxalate in the ratio of about four parts by weight magnesium oxalate with the total oxalate present in an amount to make up about 20 to 40% by weight of the solid of the fibrous slurry.

The slurry is used as a furnish in a conventional papermaking machine to produce a cigarette paper fabricated of cellulosic fibers containing 20 to 40% by weight of calcium and magnesium oxalate.

Example 2: Separate, Sequential Addition of Ashing Ingredients — Separate aqueous solutions are prepared containing the calcium cation in one solution in an amount within the range of 1 to 15% by weight and preferably 3 to 7% by weight, while the other aqueous solution contains the oxalic acid anion in equivalent amounts. The two solutions are admixed immediately prior to application with the cellulosic material or the separate solutions are applied substantially simultaneously to the cellulosic material or paper web, followed by drying. Reapplication can be made a number of times until the desired concentration of ashing ingredients is inserted within the cellulosic system.

Example 3: Oxalate Formed Internally Within Pulp Fibers — Purified cellulose pulp is slurried with 25 times its weight of a freshly prepared mixture of equal parts of 5% solution of calcium acetate and a 5% solution of oxalic acid. The slurry is cast on a screen in the usual manner for the formation of cigarette paper. The calcium acetate and oxalic acid do not immediately react to form an insoluble calcium oxalate, especially when the slurry is maintained at ambient temperature or below. As a result, the components are capable of soaking into the cellulose fibers before the calcium oxalate is precipitated to provide internal loading in situ in the fibers with the mineralizing agent. This pulp is then dried, the process repeated a number of times until the desired ash level is achieved.

Example 4: Selective Oxidation of Previously Formed Cigarette Paper – Cigarette paper, previously formed of purified woodpulp, is immersed in 100 times its weight of liquid nitrogen dioxide and maintained at a temperature of 20°C until 50% of the C_6 methylol groups have been oxidized to carboxyl groups (about 2 days). During this period, air is bubbled through the reaction vessel at a rate of 5 to 50 parts by weight per part by weight of cellulosic material, spread over the reaction time. The oxidized cellulosic material still retains its fibrous characteristics such that the paper still has strength and integrity and can be passed through two to three water washes wherein deionized water is flushed through the oxidized paper. The paper is dried to about 40 to 50% relative humidity and calendered for subsequent use in the conventional manner as cigarette paper.

Example 5: Selective Oxidation of Woodpulp – Purified woodpulp is introduced into a pressure vessel with 50 times its weight of liquid nitrogen dioxide containing 1.5% by weight water. The mixture is maintained under a constant state of agitation by recirculating a small stream of the liquid oxidizing medium from the bottom to the top of the reaction vessel. The reactants are maintained at a temperature of about 35° to 40°C under autogenous pressure and reaction is continued until 40 to 60% of the methylol groups on the C_6 position have been oxidized (about 2 hours). When reaction is completed, the liquid oxidizing medium is drawn off and the oxycellulose is washed several times with water.

The oxycellulose can be suspended in aqueous medium to form a slurry which can be employed in the conventional manner to form cigarette paper since the oxycellulose still retains sufficient of its fibrous characteristics for interfelting by the paper forming machine to produce a cigarette paper which can be processed and used in the conventional manner.

Example 6: Reduction of Oxycellulose Material – Cellulosic paper fibers previously oxidized are immersed in a 0.2% water solution of sodium borohydride in which the pH has been adjusted with oxalic acid to within the range of 6.5 to 7. Exposure in the ratio of about one part by weight oxycellulose to 100 parts by weight of solution is continued for from 1 to 10 minutes. Thereafter, the solution is drained from the oxidized cellulosic material and the treated fibers are rinsed with one or more increments of deionized water.

Aluminized Paper for Environmental Test Kit

H.L. Stroterhoff; U.S. Patent 3,874,995; April 1, 1975; assigned to the U.S. Secretary of the Army, describes a method of producing a material for chemical spot test systems containing multiple spot tests enclosed in a container which can be carried in a jacket or shirt pocket. The spot tests are designed to detect chemical species down to 0.02 part per million in water and air and are contained in packaging and support means made of weldable and impermeable plastic film.

The use of aluminized paper in the apparatus overcomes the previously encountered aluminum disc leakage problem, because $CuCl_2$ is absorbed by the paper and excess $CuCl_2$ is thereby held within the apparatus. By dispersing aluminum particles through paper, a larger surface area is presented to be contacted by and reacted with the $CuCl_2$ solution to produce instant heat and overcome the problem of difficult wetting of the aluminum discs by the $CuCl_2$ solution. Lastly, the aluminized paper is more easily handled than aluminum discs. A typical procedure for preparing the aluminized paper for the apparatus is described in the following example.

Example: Approximately 3.0 grams of conventional blotting paper, approximately 9.0 grams of 325 mesh aluminum powder, and 2,000 ml of water are mixed at high speed for approximately 3 minutes in a conventional blender, such as a Waring blender, to separate the individual fibers. The resulting slurry is emptied into a conventional paper mold, and the water is allowed to drain completely out. The resulting aluminum paper mat is removed from the paper mold screen, placed in an oven and dried at approximately 90°C for approximately 2 hours, and cut to sizes suitable for assembly into the apparatus. The thickness, weight, ratio of paper pulp to aluminum, and paper characteristics are adjustable to meet the requirements for any given application.

The portable apparatus and test is described with reference to Figure 9.5. The plastic film to produce cover **4**, package **11**, and module **6**, shown in Figures 9.5a and 9.5c was produced by subjecting a commercially available chlorofluorocarbon film of the general formula $(CF_2{-}CFCl)_X$ to a vacuum of approximately 1 millimeter of mercury at 195°F until the volatile content of the film was reduced to no greater than 0.05 part per million when measured at 150°F under a vacuum of approximately 1 inch of mercury; the measurement of 150°F being made in the conventional manner. The resulting vacuum treated film remained pliable, weldable and impermeable. Configurations **4, 11** and **6** are stamped from the above described plastic film by conventional machine shop stamping technique.

A system for detection of nerve agent presence in air or water is provided by inserting sintered glass disc **7** in recess **1** and superimposing an enzyme spot disc **8** on disc **7**, the enzyme spot being any conventional composition such as that impregnated in the enzyme paper described in U.S. Patent 3,049,411; inserting a buffer in a plastic film package **9**, the buffer being a composition as described in the aforementioned U.S. patent, in recess **10**; inserting aluminized paper **12**, manufactured as described, in recess **2**; inserting $CuCl_2$ in a plastic film package **13** in recess **14**; and inserting, in recess **3**, a substrate **15** of composition such as described in U.S. Patent 3,515,644.

A system for detecting mustard agent in air or water can be provided in combination with the above described nerve agent detection system by inserting a plastic film disc **16**, made of plastic film, in recess **17** of module **6** and superimposing on disc **16** a mustard agent spot disc **18**, containing 4-nitrobenzylpyridine impregnated in a conventional substrate; inserting package **11**, made of plastic film, containing K_2CO_3 within recess **19**; and superimposing module **6** on section **5** of the container shown at **20** in Figure 9.5c.

The plastic film used in the apparatus has particular utility regarding the mustard agent detection system, because this system is very sensitive to contamination by volatile matter within commercially available plastic film. After assembly of the nerve agent system components and mustard agent system components as described above; container **20** with module **6** superimposed thereon, cover **4** superimposed on container **20** and module **6**, and aluminum shield **21** superimposed on cover **4** are laminated together utilizing conventional laminating technique to form the integral chemical system unit shown at **22** in Figure 9.5e.

FIGURE 9.5: ALUMINIZED PAPER FOR ENVIRONMENTAL TEST KIT

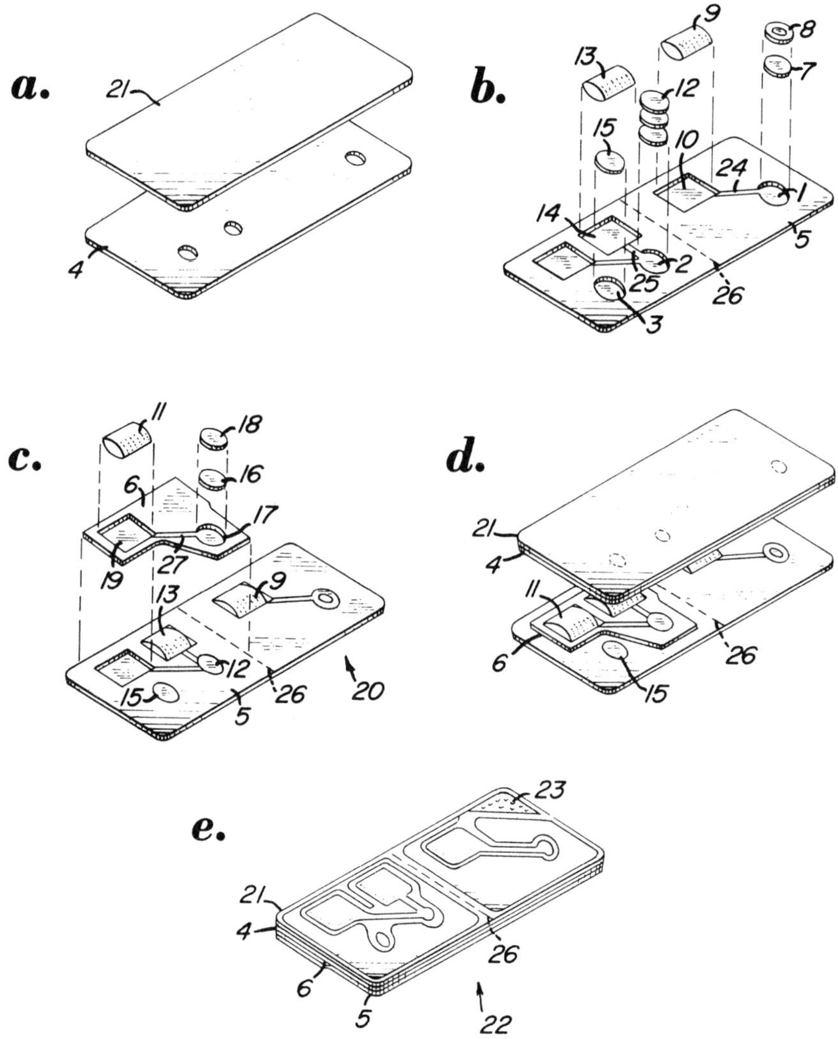

(a) Aluminum Protective Shield of Plastic Film
(b) Exploded View of Components to Be Assembled Within Plastic Container Means
(c) Components to Be Assembled Within Module Means
(d) Structure of (a) After Lamination, Preparatory to Lamination of Entire System
(e) Entire Assembly of Structures (a) Through (d) After Lamination

Source: U.S. Patent 3,874,995

The operation of the above described nerve agent and mustard agent detection system is as follows. To test the environment for nerve agent presence, shield **21** is peeled off by means of serrulated corner **23**, as shown in Figure 9.5e, and the shield discarded; package **9** is ruptured by squeezing to cause the buffer solution content to flow through channel **24** to thoroughly wet disc **8**; unit **22** is exposed to the environment containing suspected nerve agent and/or mustard agent for approximately six minutes; package **13** is ruptured and the $CuCl_2$ solution squeezed along channel **25** to contact aluminum paper **12** to generate sufficient heat to achieve adequate sensitivity for the mustard agent spot test, completion of the mustard agent test being subsequently described; unit **22** is folded along line **26** and spot disc **8** held against substrate **15** for approximately two minutes; upon folding unit **22** back to the position shown in Figure 9.5e, no color on spot disc **8** indicates presence of nerve agent which has inhibited the enzyme action, and, if no agent is present, a blue color results in disc **8** as a result of the substrate-enzyme reaction.

To complete the test of the environment for mustard agent presence, after all of the steps described above regarding nerve agent detection have been performed, package **11** is ruptured and the K_2CO_3 solution squeezed along channel **27** to wet spot **18**; no color change in spot **18** after wetting by K_2CO_3 indicates no mustard agent is present, and a change to a purple color, after spot **18** is wetted by K_2CO_3, indicates the presence of mustard agent.

Imparting Woven or Knitted Fabric Properties to Paper

The process developed by *Y. Fukumara; U.S. Patent 3,511,751; May 12, 1970; assigned to The Toyo Rubber Industry Co., Ltd, Japan* comprises chemically treating a paper to loosen the combinations among the fibers forming the paper and then pressing a woven or knitted fabric against the surface of the paper to cause orientation of the fibers forming the paper in regular directions.

The characteristics of textile fabric are imparted to paper by pressing a textile fabric to a paper which has been treated to convert the unoriented, immobile structure of the fibers to a mobile state so as to move and orient the fibers in parallel and perpendicular directions to the directions of the yarns in the textile fabric.

The method comprises soaking a paper in an aqueous alkali to convert the cellulose fibers forming the paper to alkali cellulose, treating the alkali cellulose with carbon disulfide to convert it to cellulose xanthate in the form of paper, thereby, to make the fibers mobile with respect to each other, pressing a textile fabric, as a forming die, against the surface of the cellulose xanthate sheet, and coagulating the so embossed cellulose xanthate sheet, and regenerating the cellulose in an acidic coagulating bath. The products obtained by the method may be employed in the fields in which textile fabrics have been employed as they have the properties of textile fabrics. In addition, they may be used in fields where papers have usually been employed. Thus, the products have great versatility.

Example: A kraft paper was soaked in an 8% aqueous caustic soda solution for 1 minute and the sheet of alkali cellulose thus formed was taken out of the aqueous caustic soda solution and squeezed to remove excess aqueous caustic alkali. The sheet was then soaked in carbon disulfide at 25°C for 30 minutes, squeezed to remove excess carbon disulfide, and interposed between two sheets of a plain cotton fabric. The assembly so formed was compressed at a pressure of 5 kg/cm^2 for 10 seconds. After removal of the cotton fabric, the sheet of cellulose xanthrate was soaked in a coagulating bath containing sulfuric acid in a concentration of 120 g/l for 3 minutes to coagulate and regenerate the cellulose and was then washed with water and dried.

The properties of the converted paper thus obtained are summarized in the table together with those of the untreated paper.

	Untreated	Treated
Thickness (mm)	0.082	0.221
Areal weight (g/sq m)	53.2	
Decrease in area (%)		17.4
Loss in weight (%)		1.6
Tensile strength (kg/sq mm)		
MD	3.78	2.23
TD	2.28	2.15
Elongation at break (%):		
MD	2.5	20.3
TD	2.1	19.8
Wet strength (kg/sq mm):		
MD	0.02	0.8
TD	0.01	0.7
Folding endurance (times):		
MD	197	100,000
TD	146	100,000
Wrinkle resistance (deg.):		
MD	30	116
TD	55	116
Gurley stiffness (mg)	150	52
Custom abrasion (times)	300	3,000
Shrinkage (%)	0.03	0.05

In the above data, tensile strength and elongation at break were measured on test pieces of 15 mm width at a rate of drawing of 45 mm/min. The tensile strength was calculated on the basis of the thickness at the thickest portions. Folding endurance was measured by means of a Schopper folding tester. Wrinkle resistance and shrinkage were measured in accordance with JIS (Japanese Industrial Standard) L-1006 (test method for woolen fabrics).

As indicated in the data, the treated paper had remarkably altered properties compared with the untreated one and there was obtained a further improvement in softness by post-treating the treated paper with a surface active agent. For instance, by soaking the treated paper in a 1% aqueous solution of Softex KW at 40°C, for 3 minutes and subsequently drying, the wrinkle resistance was improved from 116° to 145°. There was obtained a further improvement in wrinkle resistance, i.e., from 116° to 165°, by minimal post-treatment with 10% aqueous glycerol. By employing an aqueous solution of glycerol with added polyvinyl alcohol, tensile strength was also enhanced. For instance, by post-treatment in which a treated paper was soaked in an aqueous solution containing 4% polyvinyl alcohol (degree of saponification = 88%, average molecular weight = 1725) and 4% glycerol for 3 minutes and then dried, the tensile strength, elongation at break and wrinkle resistance were improved compared with those of the original treated paper from 2.2 kg/sq mm to 2.8 kg/sq mm, from 20% to 25% and from 116° to 165°, respectively.

INVENTOR INDEX

U. S. PATENT NUMBER INDEX

COMPANY INDEX

The company names listed below are given exactly as they appear in the patents, despite name changes, mergers and acquisitions which have, at times, resulted in the revision of a company name.

NOTICE